D1827288

Historical & Cultural Astronomy

The Historical & Cultural Astronomy series includes high-level monographs and edited volumes covering a broad range of subjects in the history of astronomy, including interdisciplinary contributions from historians, sociologists, horologists, archaeologists, and other humanities fields. The authors are distinguished specialists in their fields of expertise. Each title is carefully supervised and aims to provide an in-depth understanding by offering detailed research. Rather than focusing on the scientific findings alone, these volumes explain the context of astronomical and space science progress from the pre-modern world to the future. The interdisciplinary Historical & Cultural Astronomy series offers a home for books addressing astronomical progress from a humanities perspective, encompassing the influence of religion, politics, social movements, and more on the growth of astronomical knowledge over the centuries.

The Historical & Cultural Astronomy Series Editors are: Wayne Orchiston, Marc Rothenberg, and Cliff Cunningham.

Thierry Montmerle • Yi Zhou

China and the International Astronomical Union

Divorce, Separation and Reconciliation (1958–1982)

 Springer

Thierry Montmerle
Institut d'Astrophysique de Paris
Paris, France

Yi Zhou
Radio France Internationale (RFI)
Issy-les-Moulineaux, France

ISSN 2509-310X   ISSN 2509-3118   (electronic)
Historical & Cultural Astronomy
ISBN 978-3-031-01789-6   ISBN 978-3-031-01787-2   (eBook)
https://doi.org/10.1007/978-3-031-01787-2

Cover illustration: A group of radioastronomers from Peking Observatory (now the Headquarters of the National Astronomical Observatories of Chinese Academy of Sciences, Beijing), in front of one of the 6-m dishes built as part of the Miyun station interferometer in 1973. The head of the group was Wang Shouguan (center). This image was first published in Sky & Telescope (March 1974 issue, p.152), in an article entitled "Astronomy in China", by George Miley, from Leiden Observatory in the Netherlands. See Sect. 5.5. (Photo G. Miley; digitized original courtesy of Sky & Telescope; with permission from Wang Ying, Wang Shouguan's daughter).

This Springer imprint is published by the registered company Springer Nature Switzerland AG
The registered company address is: Gewerbestrasse 11, 6330 Cham, Switzerland

Prologue

By the standards of a young country like mine, astronomy has a long history of international collaborations, going back at least to the "Celestial Police" of Baron Janoš von Zach (1754–1832) who, in 1798–1799, asked his European colleagues to search their collective skies for a missing planet between the orbits of Mars and Jupiter (according to the Titius-Bode scheme). Later came the *Carte du Ciel* program, promoted by Admiral Ernest Mouchez (1821–1892), Director of the Paris Observatory, who had earlier had only moderate luck in trying to coordinate expeditions to observe the 1882 transit of Venus. Still later came the International Union for Coordination of Solar Research (1904–1913), the brainchild of George Ellery Hale (1868–1938).

Of course, these are all very recent by Chinese standards, given that astronomers–astrologers there were recording useful data on comets, meteor showers, sunspots, and transient events at least two thousand years before Mouchez assembled his colleagues in 1887 to figure out how to photograph and catalog the entire sky. For the history of Chinese astronomy, there are many serious sources. But if you have not read Joseph Needham at length, I recommend Joseph Needham in 11 pages (1962 Quarterly Journal of the Royal Astronomical Society, vol. 3, pp. 87–98). European astronomy came to China in the hands of Jesuits, at a time when they were required to teach and believe in a geocentric universe. China was, in any case, accustomed to magnificent bronze equipment suited for this point of view: astrolabes, armillary spheres, gnomons, and so forth.

This brings us to a war, not otherwise mentioned in the following pages: the so-called Boxer Rebellion of 1899–1901. This was put down with the assistance of the Eight-Nation Alliance. The Eight Nations were Great Britain, France, Germany, the USA, Russia, Japan, Italy, and Austria-Hungary. In the process, soldiers from at least some of the Eight Nations looted all sorts of books, manuscripts, art, and statuary from the parts of China they occupied, including instruments from what is now the Old Beijing Observatory. The Boxer Protocol between China and the Eight Nations was signed on September 7, 1901, and seems to have been one of the world's truly unequal treaties. In it, China was expected to send 450 Taels of fine silver (about 18,000 tons) over 39 years, 28.97% to Russia, 7.32% to the USA, and so forth. Payments began apparently in gold, but in due course some of the Eight began to feel the injustice of it all. (Germany gave up her claims in 1919, bound by another treaty that we will meet in a moment.)

Why are we wandering in this territory? Because, beginning in 1909 the USA, under Theodore Roosevelt's administration, remitted its share to China (and France and Britain did similar things about the same time), with the understanding that the $17,000,000 still owed to the USA would be used to pay for scholarships for Chinese students to come study at American Universities. And Chang Yu-Che (1902–1986; now "Zhang Yuzhe" in the modern, pinyin transcription), perhaps the single most important character in the drama which follows, came to Otto Struve's Yerkes Observatory with one of those scholarships, after receiving his first degree in China in 1923, to earn a University of Chicago PhD in 1929 with a thesis on orbits of binary stars. He returned to Yerkes in 1947–1948, when Adriaan Blaauw was also there; they can be identified in a 1947 staff photo (Fig. 7.11 in the book; by the way, my thesis adviser, Guido Münch, can be seen in the upper left corner of that same photo).

About 1300 Chinese students came to the USA between 1909 and 1929 with Boxer Indemnity scholarships, including at least two you might have heard of, Chen Ning Yang (who, with Tsung-Dao Lee, developed the theory of parity non-conservation, confirmed by experiments organized by Chien-Shiung Wu, a female physicist who did not receive such a scholarship, nor share their 1957 Nobel Prize in Physics), and Chia-Chiao Lin, who with his student Frank Shu developed the density-wave theory for spiral structure in galaxies. (All of them settled in America.)

And then there was another war, called The Great War, until its greater successor demoted it to World War I. This is an essential part of our story as you already know, because it led fairly directly to the founding of the International Astronomical Union. But there is another connection, made clear in the Treaty of Peace, signed at Versailles on June 28, 1919. I shall tell you a bit about it, in case you don't have a copy at hand. Mine is on the table next to this desk and was purchased at an auction a decade or so ago, for a sizable sum, which I have never regretted. There is a list of a couple of dozen international agreements, conventions, etc., which were allowed to continue to exist, mostly about health, customs, standards, etc., but definitely not the *Carte du Ciel* (though it lingered as Commission 23 of the IAU for years), or Hale's Solar Union.

Indeed, there is only one mention of astronomy in the entire 2-in., thick, $13 \times 8\frac{1}{2}''$ Treaty document. It is Article 131, in the middle of seven relevant to China. The contents are, roughly:

- 128: "*Germany renounces all benefits and privileges under the Protocol signed September 7, 1901, and renounces in favor of China any reparations under the Protocol due after 1917 March 14*" (China would join the Allied Powers in World War I on August 14).
- 129: is about trade and customs, mostly about not giving Germany any advantages.
- 130: gives back a number of buildings, wharves, munitions, and so forth to China, but reserves ownership of legation buildings and some others.
- 131: "*Within 12 months of the coming into effect of the Treaty, Germany undertakes to restore all the astronomical instruments which her troops removed in 1900–1901, paying for the packing, shipping, and re-installation in Peking at German expense.*" (The items, which did not include telescopes, were actually restored some time around 1926.)

- 132–133: More about land and property.
- 133: Germany renounces to Britain and to France and China various items in the British and French concession areas.

Who signed for China et Versailles? Wikipedia doesn't mention it. My copy of the Treaty says: Mr. Lou Tseng Tsiang, Minister of Foreign Affairs, and Mr. Chengtian Thomas Wang, former Minister of Agriculture and Commerce (along with the American President Woodrow Wilson, and, for Poland, the famous pianist Ignacy J. Paderewski, who was President of the Council of Ministers, and many other fascinating folks).

At this point, I temporarily hand you over to the main text here to learn about China's first adherence to the IAU in 1935.

We come back as war clouds are again darkening, especially in the parts of northern China closest to Japan. Zee zhen (Zhen-Zee)'s 1941 paper in the Japanese Journal of Astronomy and Geophysics on an asteroid orbit, in which he thanks Prof. K. Hirayama (emeritus at Tokyo Imperial University) was very possibly the last such collaboration for quite a while (this was just before Pearl Harbor, when the USA entered WWII).

At this point, please stop and read Chang Yu-Che's 1942 Popular Astronomy article (vol. 52, p. 198), in which he records a cordial reception on Hokkaido for the 1936 Chinese eclipse expedition there, in contrast to the conditions under which he and his colleagues observed another eclipse in September 21, 1941, with Japanese bombers passing overhead and at least one bomb landing quite close to them. Even his Observatory was operating under internal exile in Kunming, following the Japanese invasion of 1937.

I believe the details of the numbers of Chinese astronomers belonging to the IAU and attending its General Assemblies are given in the main text. I would note only that the numbers peaked at the Stockholm in 1938; that the country managed more than once to catch up on its allotted contributions to the IAU, under what must have been very difficult circumstances; and that by 1958 (Moscow) only two astronomers from China were still members of IAU Commissions: Tcheng Mao-lin (now "Cheng Maolan," a biography of whom is currently in press with Springer by the same authors), from "Academia Sinica, Building 14, Apartment 301, Chungkuan-tsuen, Western Suburbs, Peking," from Commission 21 ("Light in the Night Sky"), and Chang Yu-Che (by then back home in Nanking), of the Purple Mountain Observatory, from Commission 38 ("Exchange of Astronomers"), founded in 1948.

So, out to the main text, and back again, Fu Chengqi and Ye Shuhua (Appendix D) having recorded that, after the 1982 Patras IAU General Assembly (where re-adherence was accomplished), the total number of Chinese individual members was 109, out of a total of 5200 for the whole IAU, or 2.1%. After 2003 (Sydney), this was 320 out of 8114, or about 3.1%: Chinese astronomy had grown somewhat faster than the IAU membership. You can find current numbers on the IAU website.

About Patras and the re-adherence: Yes, I was there and heard several speeches, recorded in the main text and in the Proceedings. But (almost?) equally notable was another breakthrough in international relations: This meeting was the first time Yakov Borisovich Zel'dovich had been allowed to travel to Western Europe, to give one of the invited discourses in that reconditioned ancient Greek theater (I knew him from an earlier meeting in Jena, East Germany, of the International Society on

General Relativity and Gravitation). It was also the last IAU General Assembly at which I had no formal responsibilities, because at a Cosmology meeting in Crete soon thereafter, Vera C. Rubin appointed me to chair a new Supernova Working Group (the previous one had died with Fritz Zwicky in 1974).

As you undoubtedly know, the 28th IAU General Assembly took place in Beijing in summer 2012, but we from other countries first had the opportunity to be welcomed by our Chinese fellow astronomers much earlier, in 1986, when two IAU Symposia happened there. No. 124 on Observational Cosmology occurred in August in Beijing, but the very first was actually No. 125 on the Origin and Evolution of Neutron Stars in May in Nanjing. I was there (#91) on the invitation of Prof. Wang Zhen-ru (1937–2017), Chair of the Nanjing University Department of astronomy. She had done a great deal of both the scientific and local organizing, and she appears on the front row (#8) of the conference photograph (© IAU).

The IAU Symposium No. 125 also provided the last opportunity for a meeting between former IAU President Adriaan Blaauw (1914–2020), and Prof. Chang Yu-Che (1902–1986), who died about two months later. This allows me to give almost the last word here to the latter, with his 1945 biography of the astronomer Chang-Heng, a contemporary of Ptolemy (Popular Astronomy, vol. 53, pp. 122–126). The paper was submitted in December 1944 from the Observatory, where he was still in internal exile in Kunming.

The good news is that Chang-Heng (or Zhang Heng, 78–139 CE) has a 2.5-page article in the Second Edition of the Biographical Encyclopedia of Astronomers (2014, Springer, pp. 2422–2424), written by Thomas Hockey, Editor-in-Chief of the Encyclopedia. The article has many references in foreign languages, almost none of which I can read, so let me please here express my great admiration for Profs. Ye Shuhua and Thierry Montmerle (as well as his Chinese colleague Yi Zhou), who, with enormous skill and generosity, read, write, and speak my language as well as their own, and others.

University of California, Irvine, CA, USA

Virginia Trimble

Preface

The most important crisis in the post-WWII history of the International Astronomical Union (IAU) is undoubtedly "the China conflict," when China withdrew from the IAU in 1960 in protest against the admission of Taiwan by the Executive Committee the previous year, and which lasted until 1982, when she formally rejoined the IAU. The highlights of this story can be found in the reference book by Adriaan Blaauw (IAU President, 1976–1979, and a major actor of the return of China) published in 1994,[1] covering however only the first fifty years of the history of the IAU (1919–1969), but suitably extended to include the "China conflict". Still, many details remained unsettled or obscure (like the implication of the US State Department in pushing for the admission of Taiwan), or even absent, like the long period (almost a quarter of a century) between the withdrawal of China from the IAU, her admission to the United Nations in 1971 (and Taiwan being expelled), and the resumption of contacts with the West after the Cultural Revolution in 1979.

On the other hand, since the publication of Blaauw's book, new, important information has become available. While Blaauw's book mainly presented "the IAU view" as revealed by its archives (which Blaauw himself did not fully analyze since the "China conflict," important as it had been, was not the main topic of his book), "the China view" was expressed in a chapter of another book, about the history of China and the IAU, published in 2009 (in Chinese): "*Under One Starry Sky*,"[2] which was translated for the present work by one of us (YZ). The authors are two astronomers of the Shanghai Observatory, Fu Chengqi (傅承启) and Ye Shuhua (叶叔华). Born in 1927, Prof. Ye Shuhua, who still goes regularly to her office and with whom we had many friendly exchanges, turned out to have a long relationship with the IAU. In 1958, she was a member of the Chinese delegation at the Moscow

---

[1] Blaauw, A. 1994, *History of the IAU, The Birth and First Half-Century of the International Astronomical Union* (Dordrecht: Kluwer Academic Publishers)

[2] Fu Chengqi, and Ye Shuhua 2009, *Under One Starry Sky* 《同一个星空:国际天文学联合会史》—*History of the International Astronomical Union.* [In Chinese: Shanghai Jiao Tong University Press, Shanghai]

General Assembly, when the application of Taiwan was discussed for the first time, leading to the "divorce" between China and the IAU, and she was also a member of another delegation twenty years later, at the Montreal General Assembly in 1979, when the first, decisive steps toward their "reconciliation" took place. Eventually, Ye Shuhua would become the first Chinese IAU Vice-President (1988–1994).

Yet, neither of the books by Blaauw or Fu and Ye give details about the "crossing of the desert" (i.e., the "separation") between the "divorce" and the "reconciliation," a period for which the IAU archives contain a rich documentation. Also, these authors do not mention the role of the International Council of Scientific Unions (ICSU) and its major partner, UNESCO, as well as of the other International Scientific Unions, during the "China conflict." This background became available in 1996 with the publication of "*A history of the International Council of Scientific Unions*"[3] by Frank Greenaway, which has, for ICSU, the same "bible" character as Blaauw's book has for the IAU.

It was on the occasion of the IAU Centenary (2019) that one of us (TM) took up the task of revisiting this painful episode, which emerged from an international Conference organized by TM and a science historian from Université Paris-Sud (Orsay), Danielle Fauque, and which took place in Paris in October that year.[4] The project for the present book was made possible as a result of many factors: examining the IAU archives kept by the Secretariat in Paris, thanks to the generous support of General Secretary Teresa Lago (2018–2021), having access to many Chinese documents (most of them either unpublished, even if translated, or unavailable), thanks to YZ, broadening the picture by searching information about ICSU, etc., and even raising the case of the Olympic Games, which brought unexpected inspiration for the solutions eventually found in 1979 for the mutual "reconciliation" between China and the IAU.

As a result, the present work aims at providing a new, in-depth story of the "China conflict" (also called "the China crisis"), with a much wider perspective than was possible before. What emerges, seen from "inside the IAU," is a complex, fascinating story of human relations and science diplomacy under the shadow of the geopolitics of the Cold War, and also the realization that this story is still highly relevant today. All in all, the China conflict appears to be important not only for the history of the IAU, but also for the history of contemporary China.

Paris, France   Thierry Montmerle
Issy-les-Moulineaux, France   Yi Zhou

[3]Greenaway, F. 1996, *A history of the International Council of Scientific Unions* (Cambridge: Cambridge University Press)

[4]Montmerle, T., & Fauque, D. 2022, *Astronomers as Diplomats: When the IAU builds bridges between nations,* (Cham: Springer Nature Switzerland)

Acknowledgements

This extensive study of the China–Taiwan–IAU relations spanning over twenty years could not have been possible without the help and advice of many colleagues. First and foremost, TM thanks Xiaowei Liu, from Kunming Observatory (Yunnan), whom he got to know when he was IAU Vice-President during his term, for enlightening discussions and having given access to the book by Fu Chengqi and Ye Shuhua, from Shanghai Observatory (*"Under one starry sky,"* 2009, in Chinese; their Chap. 5), which tells for the first time "the China view" of this period. This Chapter was translated by one of us (YZ) with the support of IAU General Secretary Teresa Lago (2018–2021), which is greatly appreciated. TM is also respectfully grateful to Prof. Ye Shuhua (born 1927) for useful correspondence and her kind remarks about this work, which culminated in sending him a personal copy of her book with a dedication.

TM is also grateful to George Miley from Leiden Observatory for interesting discussions, Yan Yihua (National Astronomical Observatories, Beijing) and Wang Ling (Wang Shouguan's daughter), for facilitating correspondence with her father, then in a nursing home, who unfortunately passed away in the course of this work at age 98 in Beijing. TM thanks Georges Meynet from Geneva Observatory, Paul Ho of the Academia Sinica Institute of Astronomy and Astrophysics in Taipei, and Kaz Sekiguchi from Tokyo Observatory, for their help in providing documents. TM also takes this opportunity to honor the memory of "his" President Norio Kaifu (1943–2019), whom he saw in generous action in the course of the elaboration of the East Asia Observatory.

TM is also privileged for having exchanged information with two past General Secretaries who played an important role in the "China conundrum": Kees de Jager in the Netherlands (who also passed away in the course of this work, at age 100) and George Contopoulos (born 1928) in Greece.

Back to France, TM is indebted to Danielle Fauque, his co-editor of the Springer book *"Astronomers as Diplomats,"* for discussions about the international scientific unions and for very useful comments on the manuscript. TM's warmest personal thanks also go to the IAU Secretariat: Ginette Rude, who, in spite of being retired, helped him many times to find his way in the cramped room housing the IAU archives in the Institut d'Astrophysique de Paris (TM's research institute), and Madeleine Smith-Spanier, IAU database Assistant, for her hospitality in the IAU office, and her kind and efficient help in solving questions about IAU members.

Last but not least, TM would like to thank Prof. Virginia Trimble, from UC Irvine, former IAU Vice-President, whom he met many times in the course of their careers and in IAU circles, for having kindly accepted to write an incisive Prologue to this book, and Pierre Léna, from Paris Observatory and the Académie des Sciences, for the live report of his visit to China in 1979, at the invitation of Prof. Ye Shuhua (Appendix C).

Contents

Chapter 1
Introduction and Outline

In the summer of 2012, the XXVIIIth IAU General Assembly took place in Beijing, China. The membership of the Union was approaching 11,000, of whom 2700 would be attending at least part of the events, spread over two weeks (August 20–31). The organizers of General Assemblies always make a point in inviting a high-level national official to give a welcome address to the participants during the Inaugural Ceremony.[1] In Beijing, this official was none other than Xi Jinping (习近平), then the Vice-President of China, already known to be the successor of Hu Jintao (胡锦涛), who would step down the following year, after two terms of the five-year Chinese Presidential mandate.

In the back of the stage, lined up in comfortable armchairs, were other Chinese officials, as well as Robert ("Bob") Williams, American, outgoing IAU President, and Ian Corbett, British, outgoing General Secretary. Sitting in the front row of the audience were Norio Kaifu, Japanese, incoming IAU President, and myself (French), incoming General Secretary, who would deliver his own address during the closing ceremony (Fig. 1.1). Once again, I marveled at the cultural and political diversity that the member countries of the IAU (nearly 100 by then) represented "Under One Sky." Then, a curious incident happened: Bob had just delivered his address, thanking our hosts, when, as Xi Jinping was coming to the lectern to deliver his own welcoming address, Bob dropped his spectacles. Xi graciously picked them up and handed them to him, drawing laughter and applause from the audience: The act by the future President of China of returning his spectacles to the American President of the IAU was interpreted as a highly symbolic gesture of international friendship and good will, and in fact the event made it on the next day to the front page of "*China Daily*," the official newspaper of China.[2] Thus began an exceptional General

---

[1] Ref.: Transactions Vol. XXVIII B (see the complete references for the IAU "Transactions" volumes quoted in the text at the end of this chapter).

[2] It can be seen in the General Assembly daily newspaper "*Inquiries of Heaven*" (issue of Tuesday, Aug. 28, p. 4 [https://www.iau.org/static/publications/ga_newspapers/20120803.pdf])

T. Montmerle, Y. Zhou, *China and the International Astronomical Union*, Historical & Cultural Astronomy, https://doi.org/10.1007/978-3-031-01787-2_1

Fig. 1.1 One of the authors (TM) delivering his speech as incoming General Secretary at the closing ceremony of the 2012 IAU General Assembly in Beijing. In the back (from left to right), Ian Corbett and Robert Williams (outgoing General Secretary and President, respectively), and Norio Kaifu, incoming President. (© IAU)

Assembly, with a remarkable bonus: the (re)admission of North Korea, just across the border, to the membership of the IAU.

During the preparatory phases of the Beijing meeting, when I was Assistant General Secretary, I had become familiar with the classic book by Adriaan Blaauw, published in 1994: *"History of the IAU: The Birth and First Half-Century of the International Astronomical Union,"*[3] which contains a chapter (#9) entitled: *"The China Conflict."* This chapter recalls a severe crisis within the IAU, when, in 1959, China withdrew from the IAU for 20 years as a protest against the admission of Taiwan. When the situation started to unlock in 1979, the IAU General Assembly was taking place on the campus of the University of Montreal, where I happened to be an Assistant professor at the Astronomy Department, so of course it was easy for me to attend. I remember vividly the applause of the audience when the news was communicated that, pending some minor legal issues to be settled after the General Assembly, China was (almost) back into the IAU community. I wasn't really familiar with the background, and for me, to have the two Chinese countries (or so I thought) within the IAU was wonderful—after all, at that time there were two Germanies and two Koreas, so why not "two Chinas"?

---

[3] Blaauw 1994; hereafter "Blaauw's book."

However, after reading Blaauw's Chapter, I understood that the crux of the "China conflict" was not "two Chinas, two representations" (as in the case of Germany and Korea), but "one China, two representations,"[4] and why solving this conundrum would take 20 years. Yet, Blaauw's chapter is relatively short (15 pages) and comes with a caveat in its Introduction:

> I shall rely mainly on archival documentation pertaining to the years I was President of the Union, i.e. 1976–1979 and, of course, published data. The archival data pertaining to the General Secretariat of those years have not yet been sorted out and described at this moment (early 1994). (. . .) Little will be said also about the efforts towards reconciliation by ICSU.[5]

I began to be really interested in investigating more about the "China conflict" on the occasion of the IAU Centenary (1919–2019: "IAU@100"). The closest date to celebrate this event was the XXXth General Assembly in Vienna (Austria), in 2018. The celebration included a dedicated Symposium: *Under One Sky: The IAU Centenary Symposium* ("IAU Symposium 349," or "IAUS 349").[6] Its "Part 5" included fascinating talks by Roland Wielen, from Germany (*"Germany's difficulties in becoming a member of the IAU,"* pp. 205–213), and Xiaowei Liu, from China (*"The China crisis,"* pp. 222–227). Thus came the idea of organizing an "IAU@100" international conference (in Paris) around the theme of *"Astronomers as Diplomats."*[7]

As former General Secretary of the IAU, and having an office at the Institut d'Astrophysique de Paris, I had an easy access to the IAU archives stored at the Secretariat on the second floor, with the help of the IAU archivist, Ginette Rude. In the course of examining these archives in the context of the Paris conference, I discovered several folders about China and Taiwan—the ones, presumably, that Blaauw did not use. The whole story of the "China conflict" was there, seen from "inside the IAU"! Over a hundred of well-kept documents (mostly letters, some of them handwritten, unpublished reports, etc.) were just waiting to be examined. In addition, a tool that Blaauw could not use was the so convenient electronic access to the "official" files of the IAU, either "private," like the minutes of Executive Committee meetings,[8] or "public," like the IAU "Transactions," which contain the Proceedings of each General Assembly (discourses, list of members, etc.),[9] or the newspapers published daily as the sessions take place.[10]

---

[4] In fact, for Germany it was different before 1961, but the other way around, since it had "two countries, one representation"! See Wielen (2019, 2022).

[5] Many quotations are given in the book. They are indicated by special layouts: downsized separate paragraphs (like here), or italics (when within the text).

[6] Sterken et al. (2019), referred to as "IAUS 349" in the text.

[7] See https://www.iau.org/news/announcements/detail/ann19056/, and the resulting book by Montmerle and Fauque (2022) ([MF22] for short in the list of References).

[8] https://www.iau.org/administration/executive_bodies/executive_committee/ecminutes/

[9] https://www.iau.org/publications/iau/transactions_b/

[10] https://www.iau.org/publications/iau/ga_newspapers/

More was yet to come. Not only was "the IAU view" revealed by the Archives available, but also "the China view," expressed in a chapter of a book (in Chinese) that Xiaowei Liu had used and which he kindly provided me: *"Under One Starry Sky,"* by Fu Chengqi and Ye Shuhua, published in 2009.[11] The relevant excerpt (translated from Chap. 5, Sects. 4 and 5; hereafter quoted as "Fu&Ye" for short) gives a vivid account of how the Chinese astronomers saw the admission of Taiwan by the IAU and the steps toward reconciliation.[12] For the convenience of the reader, since the text is unavailable in the West, it is reproduced in full (translated by one of us, YZ) in Appendix D.

However, neither Blaauw nor Fu&Ye give details about the "separation" period between the "divorce" and the "reconciliation." Neither do they mention the role of ICSU and its major partner, UNESCO, as well as of the other International Scientific Unions, in the "China conflict." This became more easily feasible in 1996 (thus after Blaauw's book was published) with the publication of *"A history of the International Council of Scientific Unions"*[13] by Frank Greenaway, a former Secretary General of the IUHPS (now IUHPST).[14]

We are thus now in a position to assemble in a coherent, never-told-before, story the three distinct episodes of the difficult mutual relations between China (Mainland as well as Taiwan) and the IAU, spanning a quarter of a century while having the Cold War as background, from 1958 to 1982: "divorce," "separation," and "reconciliation."

This book is thus organized as follows. Chapter 2 provides a historical background, focused on two countries that suffered a long occupation or invasions by Imperial Japan: China (Taiwan and the Mainland), and Korea. Chapter 3, the longest, explains in detail how the application of Taiwan to the IAU came about in 1958, how it was accepted, how China then decided to withdraw, and how the whole process was ratified by the IAU General Assembly in 1961. Chapter 4 broadens the picture, by comparing the situation of the IAU at the time with that of other Unions.[15] Chapter 5 describes the efforts made by ICSU and the IAU to find a compromise and convince China to rejoin the Unions and, in parallel, the initiatives taken by independent astronomers to develop contacts and collaborations with their Chinese colleagues. Chapter 6 then focuses on the pressure exerted directly by UNESCO on ICSU and the Unions to expel Taiwan, after the admission of the People's Republic of China to the United Nations in 1971. Chapter 7 describes the uncertain road to reconciliation, when after a seemingly hopeless situation, the

---

[11] Fu Chengqi and Ye Shuhua (2009). Ye Shuhua (叶叔华) would eventually become IAU Vice-President (1988–1994). Born in 1927, she was (and still is) affiliated to the Shanghai Observatory. See also below, Sect. 7.3. The reference, when labeled "Fu and Ye (2009)", refers to the original Chinese edition.

[12] The "Taiwan view" can be found in S.-Y. Wang et al. (2022); see also below, Chap. 8.

[13] Greenaway (1996). Referred to simply as "Greenaway" for short in the text.

[14] IUHPST: *International Union of History and Philosophy of Science and Technology*

[15] In the present text, "the Unions" is short for "the International Scientific Unions," and "the Union" is an alternate for "the International Astronomical Union" or for "the IAU."

process suddenly accelerated in early 1979 as a result of the diplomatic opening of China, resulting in a major improvement in its relations with the USA (at the expense of Taiwan), to end up with a final, happy settlement of the "China conflict" in 1982.

In conclusion, Chap. 8 gives my own answers to several questions raised by this twenty-year-long conflict: Was the admission of Taiwan "political"? Was it legal? Was it right? Why did the conclusion of the conflict take only a few months? The final Chap. 9 is, in form of an Epilogue, a brief account of the contemporary scientific collaboration between institutes belonging to former enemy countries that gave rise a few years ago to the foundation of the "East-Asian Observatory" (EAO), modeled after the "European Southern Observatory" (ESO), with a large partnership between observatories of China (Mainland and Taiwan), Japan, South Korea, and other regional partners.

Chapter 2
Historical and Geopolitical Background

2.1 From a World War to the Next

The IAU came into being in Brussels on July 24, 1919, as the world was just starting to see the light again after the horrors of World War I. This birth, as well as that of other international scientific unions, took place in difficult conditions, under the aegis of the International Research Council. The hope was that scientists would entertain a collaborative spirit around the globe, and help promote peace, in spite of the many political hardships their countries would inevitably face in the future. In practice, this hope was initially only partially fulfilled, since in 1919 only the victor countries (the Allied Powers) could be admitted into the various unions, the defeated countries (the Triple Alliance), notably Germany, and even neutral countries, being banned from joining. It took another 12 years, with the creation in 1931 of the International Council of Scientific Unions (ICSU) for all restrictions to be removed on country applications for membership.[1]

Unfortunately, less than a decade later, the world plunged again into a war that almost no country could avoid, not even neutral ones, with two main battlefields: all across Europe and the Mediterranean arc on the one hand; Asia and the Pacific Ocean, on the other. The IAU stopped functioning normally for 10 years (from the 1938 General Assembly in Stockholm to the first post-war General Assembly in 1948 in Zürich), although a dispersed Executive Committee continued to operate, in spite of the death of its President, Arthur S. Eddington, on November 22, 1944, at age 62. The General Secretary elected at the Stockholm General Assembly was the Dutch astronomer Jan Oort, who would play a key role in our story 20 years later as IAU President. After the invasion of his (neutral) country by German troops in May 1940, Oort had to be replaced for the duration of the war by Walter S. Adams, Director of Mount Wilson Observatory in California. But he participated in the

---

[1] D. Fauque and R. Fox (2022) give a detailed account of this period. For a summary, see Blaauw's book, Chaps. 1 & 2, and Montmerle (2019; IAUS 349).

© The Author(s), under exclusive license to Springer Nature Switzerland AG 2022
T. Montmerle, Y. Zhou, *China and the International Astronomical Union*, Historical & Cultural Astronomy, https://doi.org/10.1007/978-3-031-01787-2_2

Dutch resistance, until he resumed his position within the IAU at the Liberation. He then organized a small conference in Copenhagen in 1946 in preparation for the Zürich General Assembly, held two years later.

But the geopolitical climate after WWII was gloomy: In Europe, the Berlin blockade decided by Stalin in 1948 started just a few weeks before the Zürich meeting and lasted almost a year, to end with the partition of Germany between two states: the "Federal" Republic, and the "Democratic" Republic, and the beginning of what came to be known as "The Cold War." The physical separation and difficulties in crossing the borders between two blocs, East, dominated by the USSR, and West, under the umbrella of the USA, was dubbed "the Iron Curtain."

This time, in spite of the tense international situation, the IAU did not want to repeat the errors of the post-WWI period: On the contrary, it became very "inclusive," by admitting new member countries (and their individual members) with rather lax criteria about the extent of their astronomical activities (only a bona fide astronomy association was required, like an Astronomical Society officially recognized by an Academy of Sciences). After all, the war had obviously shattered and brought to ruins many countries; doing so was simply motivated by an optimistic bet on the future (see Table 2.1).

However, at the other end of the world, in Asia, another story unfolded, as horrific as it had been in Europe. The tragic events that took place there in the first half of the twentieth century in Asia are perhaps less well known in the West. It is well beyond the scope of this paper to describe them in any detail, but to the extent that they deeply affected the birth and development of modern astronomy in East Asia, I will present here a condensed historical background which is necessary to understand the origins of the "China crisis" that occurred within the IAU in the late 50s, and, for comparisons' sake, the situation within the IAU of the other Asian "divided country": Korea.

2.2   Historical Background in East Asia: Taiwan

At the turn of the twentieth century, China and Korea were ruled by century-old dynasties: the (Manchu) Qing dynasty in China (since 1636), and the Joseon dynasty in Korea (since 1392). But a new, rapidly modernizing neighbor appeared on the stage, after more than two centuries of almost complete isolation from the outside world: the "Meiji-restoration" Imperial Japan (since 1868).

In 1883, China declared war on France, as it was conquering Tonkin (with Hanoi as its capital) to expand her colonization of Indochina up to mountains to the North marking the Chinese border (Yunnan and Guangxi provinces). The French bombarded Formosa ("the beautiful," the name of Taiwan in Portuguese) in 1884 and isolated it by a blockade. Empress Dowager Cixi was eventually defeated in 1885, but, recognizing its strategic importance, she had time to upgrade Formosa to "a Province of China." Then, the so-called "First Sino-Japanese war" erupted in July 1894 over the respective influence of the two empires over Korea. The war ended a

Table 2.1   Adherence of National Members and Individual Members to the IAU: 1946–1961

(Source: IAU Transactions B for each General Assembly)

Year	General Assembly (GA)	General Secretary	Country[2,3]	Ind. Members[7]
1946	(Copenhagen conference)[1]	J. Oort	India[4]	7
1947			Hungary	1
1948	Zürich, Switzerland	B. Strömgren	Finland	5
1951			Germany	0[8]
1952	Rome, Italy	Th. Oosterhoff		
1953			Venezuela	0
1954			Israel	1
1954			Lebanon	1
1955	Dublin, Ireland	Th. Oosterhoff	Austria	1
1957			Bulgaria	1
1958	Moscow, USSR	D. Sadler		
1959			Taiwan[5]	2
1961	Berkeley, USA	D. Sadler	Taiwan[6]	2
			DPR Korea	0
			Turkey	2

[1] Executive Committee preparatory Conference for the first post-war GA (March 6-13, 1946). J. Oort had been IAU General Secretary since 1935 (Paris GA), to step down in 1948 (Zürich GA).
[2] Japan re-admitted in 1952; Brazil re-joined in 1961.
[3] Adherence (i.e., new members), as accepted by the Executive Committee.
[4] India became independent later (on Aug.15, 1947).
[5] Admission in 1959 by the Executive Committee (#19; Sep. 7-9, Herstmonceux Castle, UK: Sect. 3.5).
[6] Ratification by National Members voting at the Berkeley GA (see text for details). The mandatory ratification of new adherences by a vote of the General Assembly was introduced in the Statutes only in 1970 (Brighton GA). The name "Taiwan" was the one used for its membership to the IAU, never the "Republic of China".
[7] Registered individual members at date of adherence. Census given at the following General Assembly.
[8] A "zero" means that no individual member was recorded at the time of publication of the corresponding Transactions B, likely for purely administrative reasons. In the case of Germany for instance, the individual members started to be listed only for the Moscow GA in 1958 (see **Table 2**, Sect. 3.6).

few months later (in April 1895) with another defeat of the Qing government, and the signature of the Treaty of Shimonoseki. The complete independence of Korea from China was recognized, but more importantly Taiwan was ceded to Japan (along with the Liaodong Peninsula, to the northwest of the China–Korea border, and the Penghu, or "Pescadores," Islands in the middle of the Taiwan Strait) (Fig. 2.1).

Fig. 2.1 Taiwan and the Taiwan Strait. The arc-shaped, dashed lines indicate the islands that are administered by Taiwan (names with an asterisk): The main ones are the archipelagos of Quemoy (middle left of the map), and Ma-tsu Tao ("Matsu"; top of the map). These islands are only a few kilometers off the coast of Mainland China and were shelled by the People's Liberation Army during the "Taiwan Strait crisis" of 1958 (see below, Sect. 4.1). (Credit: https://www.populationdata.net/cartes/taiwan-relief/; overlays by TM)

However, in reaction to the treaty, independence from China was declared by local notables and a short-lived "Republic of Formosa" was established, to be crushed a few months later following the invasion of the island by Japanese troops.[2]

Taiwan remained under the Japanese rule for 50 years, until Imperial Japan surrendered in 1945. Then, the island was put under the control of the Republic of China (headed by the Nationalist leader Chiang Kai-shek) as a Province, with a short-lived local government called the Taiwan Provincial Administrative Office. These were difficult years, with a corrupt administration and clashes between the population and the authorities. The Nationalist government had to send troops from the Mainland and reorganize the local government. On May 16, 1947, the Taiwan Provincial Government was therefore established. But in 1949 the civil war between the Nationalists and the Communists (led by Mao Zedong), who had been (temporary) allies during the war against Japan, turned into a defeat for Chiang Kai-shek, who fled to Taiwan and installed what remained of the Republic of China in Taipei. About 2 million immigrants crossed the roughly 130-km wide Taiwan Strait to settle in the island. They were welcomed at first, but soon felt overwhelming since the island's population was only 8 million.

Initially, the Truman administration, which had no esteem for Chiang, seemed prepared to recognize Communist China. But on February 14, 1950, the new "People's Republic of China" signed with the Soviets a "Sino-Soviet Treaty of Friendship, Alliance and Mutual Assistance." This meant that China had decided to side entirely with the USSR, which made possible the Chinese intervention in the Korean War (see next subsection). A mutual defense treaty was signed between the Republic of China and the USA in 1955, under the Eisenhower administration, and the USA maintained a military force on the island until 1979, while protecting the Strait with the mighty Seventh Fleet.[3]

2.3 Korea

As for Korea, after the Treaty of Shimonoseki Imperial Japan continued its expansionist policy to become the dominating regional power. After having defeated China in 1895, Japan waged war in 1904, this time against Tsarist Russia. The powerful Russian Baltic fleet, which had come all the way from its base in Kronstadt via the Cape of Good Hope, was defeated by the Japanese Imperial navy in the Strait of Tsushima in 1905, and to protect Japan's interests in the region from the influence of Western powers, a treaty was signed that same year between Korea and Japan,

---

[2]The population of Taiwan was 3 million in 1906 (over 96% being descendants of immigrants from coastal provinces of Mainland China, the rest being descendants of the first inhabitants having settled in the island about 6000 years ago). It had doubled by 1945.

[3]Sect. 3.7. For a contextual background about China and Taiwan, see, e.g., the books by Short (1999), and Roux (2016), respectively.

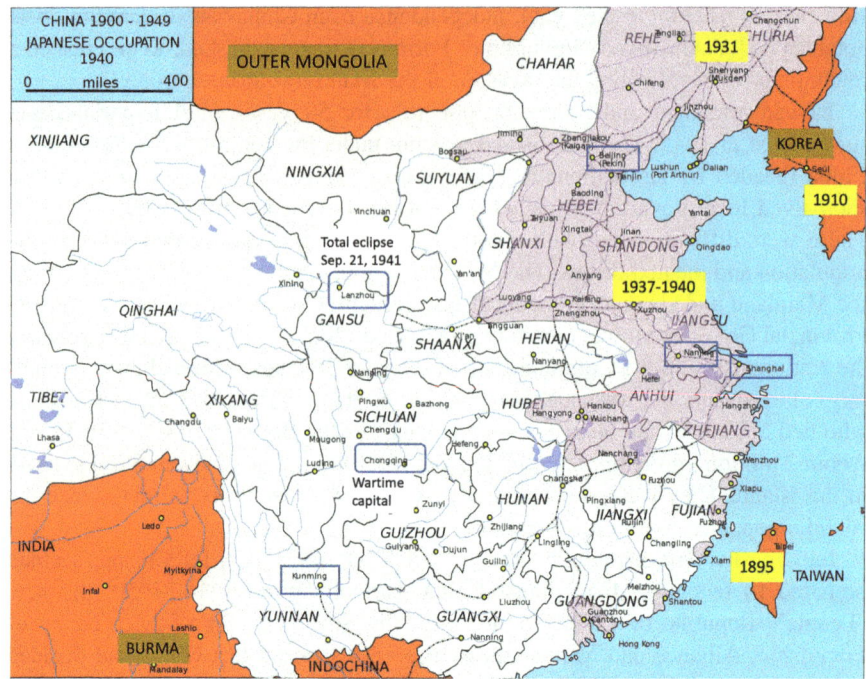

Fig. 2.2 Japanese occupation of East Asia, 1895–1940 (Dept. of History, U.S. Military Academy). Rectangular labels: cities with observatories. The Chinese astronomer Y.C. Chang, then relocated in Kunming (see Sect. 2.4), put together an expedition in 1941 to observe a solar eclipse near Lanchou (Lanzhou) (See Y.C. Chang's biography [Zhou 2022], and the *"Notes on Chinese names"* below). Rounded labels: cities mentioned in the text. (Credit: https://upload.wikimedia.org/wikipedia/commons/thumb/c/cb/Japanese_Occupation_of_China_1940.svg/2560px-Japanese_Occupation_of_China_1940.svg.png) (Overlaying labels by TM)

establishing Korea as a protectorate. Japan increasingly took control of the country, triggering resistance acts by Korean nationalists. The assassination of the Japanese Resident-General in 1909 led to the full annexation of the Korean Empire in 1910 (Fig. 2.2).

Imperial Japan had ruled over Korea for 35 years, until its surrender of 1945. Being without a government, it had been agreed at the Yalta conference (February 4–11, 1945) that Korea would be temporarily administered by the Allies, i.e., in that part of the world, by the USSR and the USA. The USSR invaded Japanese-occupied Manchuria (the "Manchukuo") on August 9, 1945, three days after Hiroshima and just prior to Nagasaki (it had signed a neutrality pact with Japan in 1941, which had been in force throughout WWII until then). Following the Yalta agreement, after the Japanese surrender the Soviet troops crossed the Manchuria–Korea border and occupied the country North of the 38th parallel, and American troops, coming from Japan, occupied the Southern part.

The idea was that some form of government could be established by democratic means in a unified country, but Soviet-American negotiations to that effect failed, and the Korean question was put to the United Nations. Elections were discussed, but for lack of an agreement with the USSR they were organized in 1948 only in the South, under UN supervision, resulting in the foundation of the "Republic of Korea." In the North, the leader of the resistance against the Japanese, Kim Il-sung, established its own government, under the umbrella of the USSR, and simultaneously created the "Democratic People's Republic of Korea." On June 25, 1950, Northern troops attacked the South by surprise with the idea of unifying the Korean peninsula under the Communist rule; they quickly invaded almost the whole country (leaving only the regions around Busan—Pusan at the time—untouched), but retreated when a US-led UN coalition pushed them back. Then China came to the rescue of Kim, and the front eventually stabilized around the 38th parallel: In the course of the war, Chinese and American troops came to fight each other face to face (and in the air, in the so-called "Mig Alley" along the Yalu river between North Korea and China, Soviet fighter pilots fought—secretly—the US Air Force).[4]

An armistice was signed in Panmunjeom in 1953, but a formal peace treaty was never signed, so "technically" the two Koreas are still at war. Nevertheless, as explained elsewhere in this volume, as far as the IAU is concerned, North Korea and South Korea are both members today, pushing for collaboration in astronomy in spite of complex, changing political relations.[5]

2.4 China

After WWI, Imperial Japan carried on its expansionist policy. In search of natural resources and labor after the 1929 economic crisis, Japanese troops engineered the Mukden Incident and invaded the northeastern part of China (Manchuria) in 1931. The occupation and colonization of this Chinese region, larger than Japan itself, was quickly followed by the establishment of a Japanese-controlled state, called "Manchukuo." This loss for China was all the more significant for the Chinese that the Qing dynasty of Emperors, which had ruled China since 1650 until the Xinhua Revolution of 1911, was of Manchu origin. In July 1937, Japan pushed its advantage over China and launched a full-scale invasion, known as "the Second Sino-Japanese war." Soon, Peking, Nanking (the capital at that time) and Shanghai fell, in apocalyptic conditions. The Nationalist government of the Republic of China, headed by Generalissimo Chiang Kai-shek, retreated in the fall of 1938 deeply into the Chinese interior, to Chungking, promoted wartime capital. But in parallel, Chiang had become a staunch enemy of the Communists, who had to endure a severe military repression (leading to the "Long March" retreat of the Red Army, led by Mao

---

[4] For more details about the Korean War, see, e.g., Cuming (2010).

[5] See Lee, H.M. (2022).

Zedong, October 1934–October 1935). For Chiang, crushing the Communists was the first priority, before resisting the far superior Japanese armed forces. But now, faced with the rapid advance of the Japanese invasion (Fig. 2.2), and as a result of the Xi'an Incident, Chiang was forced to change his policy with regard to the Communists. The two sides were able to arrive at a truce, and eventually joined forces in an all-out resistance against the invaders, under the command of Chiang (at least on paper). Outside help was minimal. The USSR was still present but had signed a neutrality pact with Japan on April 13, 1941, and was busy fighting back the sudden German invasion of June 22, 1941 ("Operation Barbarossa"), and the USA was discreetly sending military material, airplanes, and pilots, but were officially still in the isolationist camp. After the Japanese attack on Pearl Harbor, on December 6 of the same year, the USA at last entered the war, and supplies by airlift from Assam (East India) to Kunming, flying above the Himalayas, began to reach the Chinese troops and American air bases in southern China. (The airlift, called "The Hump," lasted until after the end of the war.)[6]

For Chinese astronomers at that time, like Y.C. Chang, the period was understandably very difficult: All the existing observatories, in Peking, Nanking, and Shanghai, were closed and inaccessible. The astronomers re-established themselves, the best they could, in Chungking. Y.C. Chang nevertheless heroically managed to organize a makeshift expedition to Lanchou (Lanzhou), in the Gansu province, in 1941, to observe a total solar eclipse "under the shadow of Japanese bombers."[7]

Eventually, the war ended, like everywhere else in the region, with the surrender of Imperial Japan in 1945. Onetime allies against their common enemy, Chiang and Mao soon entered a new Civil War to seize power and unify China. However, neither reached that goal: While, on October 1, 1949, Mao established the "People's Republic of China" in Peking, his new capital, as the leader of 400 million Chinese, the defeated Chiang managed to retreat to Taiwan, the new incarnation of the "Republic of China" (advertised as "Free China" for the Western public), with Taipei as its capital. China had not been united: it had become (and still is) a "divided country."

Most of the left-leaning Chinese intellectual elite, like astronomers (including Y.C. Chang), as well as those who became disillusioned with Chiang's government during the Civil War, sided with the Communists or simply stayed on the Mainland, but many notable scholars chose to flee to Taiwan or Hong Kong. On the Mainland, the Observatories progressively reopened. But state institutions like the Academia Sinica moved to Taipei.

---

[6] On the isolationism of the USA during this period, see, e.g., Olson (2014); and on F.D. Roosevelt's motivations for creating and maintaining "The Hump", see, e.g., Plating (2011).

[7] Zhou (2022). Y.C. Chang reported his expedition in *Popular Astronomy,* Vol. 50, p. 198, April 1942.

2.5 Note on Chinese Names

The transcription from Chinese proper names (personal or geographical) is always a problem when considering modern historical periods. The so-called "hanyu pinyin transcription" (meaning "to assemble the sounds of the Han language") has been adopted by the Chinese government in 1979, to replace the western-style spelling (so-called Wade-Gilles, 1912), which itself was in practice implemented differently depending on the languages considered. The idea behind a transcription elaborated by Chinese scholars, and not abroad, besides being a quite natural one, was to enforce the *same spelling* for Western languages (while only roughly solving pronunciation problems). The pinyin transcription was also adopted in Taiwan in 2009, but enforced for official documents only.[8]

For personal names, the Chinese tradition is to put the family name first, and the given name second (which usually includes one or two ideograms), but its implementation in the nineteenth and twentieth centuries was most often "westernized," i. e., used in the reverse order, although not in a systematic way. Whatever the order, it is always preferable to write Chinese names in full, since their transcription may change depending on the period considered. This is particularly true for the present work, which covers the transition between the old "western" and the new, post-1979 pinyin transcription periods.

In addition, to reflect Chinese names accurately, we also need to consider the fact that there are two prevalent ways of Chinese writing today: the so-called Simplified Chinese and Traditional Chinese. The former is used in Mainland China and the latter is used in Taiwan. Therefore, we have included both forms of ideograms as applicable in this book when writing names of important characters in Chinese.

To illustrate these important points, let's take the example of the astronomer "Y. C. Chang" (1902–1986), whom we have already mentioned in this Section and whom we will cite very often: He was the President of the Chinese Astronomical Society from 1943 to 1985 and was thus the key contact person for the IAU in China throughout the full period of the "divorce and reconciliation."

"Y.C. Chang" is in full "Yu-Che Chang," sometimes written "Chang Yu-che" in the Chinese fashion. It has become "Zhang Yuzhe" (张钰哲) in the modern hanyu pinyin transcription (family name first: Zhang [张], given name second: Yuzhe [钰哲]). Since all the documents of this period found in the IAU archives (except sometimes after 1979), including exchanges of correspondence with him, etc., use the western-style form "Y.C. Chang," or Yu-Che Chang, we prefer not to use the pinyin transcription, but stick to "Y.C. Chang" (or sometimes "Chang" to lighten the text). However, modern Chinese sources (like the book by Fu Chengqi & Yu Shuhua 2009) use the pinyin transcription, so it may happen that, in the case of quotations for example, both transcriptions co-exist in the same parts of the text. Nevertheless, once the correspondence is given, the writing is easy to follow.

---

[8] Source : Wikipedia (https://fr.wikipedia.org/wiki/Hanyu_pinyin)

An additional remark concerns Taiwan. The country did not follow the pinyin transcription of names introduced by the PRC, but sticks to the Wade-Giles system and the Western order in force before the pinyin reform, the names in Chinese being of course unchanged. For instance, the Taiwanese Nobel Laureate name 李遠哲, which is "Li Yuanzhe" in pinyin, following the order of the ideograms ("Li" family name: 李, "Yuanzhe" first name: 遠哲), is "Yuan Tseh Lee" (then abbreviated Y.T. Lee) in Taiwan. (The same rule applies to Chinese names in the USA, with the possible addition, or replacement, of the first name by a Christian name.)

For geographical names, we use throughout the text the "old" spelling as it appears in the IAU archival documents: "Peking" (sometimes even the French "Pekin"), rather than the modern "Beijing," "Nanking" instead of "Nanjing," etc., giving however the modern transcription (and the name in Chinese) wherever appropriate.

Chapter 3
The Divorce

3.1 From Dublin to Moscow: Taiwan Appears on the Scene

On August 28, 1955, the Business Meeting of the IXth IAU General Assembly was held in the Mansion House at Dublin, chaired by Otto Struve, outgoing President of the IAU. Item 11 on the agenda was the "Determination of the place of meeting for the tenth General Assembly." B.V. Kukarkin, the official representative of the USSR, addressed the Assembly to convey the invitation of the Academy of Sciences of the USSR to hold the Xth General Assembly in Moscow. The IAU Transactions[1] record that the invitation was accepted "by acclamation." A few minutes later, after having noted that no decision could be taken at such an early stage,[2] O. Struve invited J.J. Nassau, chair of the American Delegation, to address in turn the Assembly, saying:

> The American Delegation has been authorized to extend, on behalf of the United States Government, an invitation to the International Astronomical Union to hold its 1961 General Assembly in the United States. It is anticipated that such a meeting will be held in Pasadena, under the auspices of the Mount Wilson and Palomar Observatories.

This moment can be considered as doubly symbolic: On the one hand, in 1958 and 1961, two General Assemblies would be hosted in succession by countries separated by the Cold War (but not under the shadow of Stalin anymore); on the other hand, Otto Struve (1897–1963), descendant of three generations of famous German astronomers, was a Russian-born émigré (as a former young officer in the Tsarist Army during WWI) who moved to the USA in 1921 and who had been in touch with Soviet astronomers when he was Director of Yerkes Observatory as early

---

[1] Transactions Vol. IX, p. 25.

[2] Nowadays, the venues of General Assemblies are selected by the Executive Committee up to six years in advance.

© The Author(s), under exclusive license to Springer Nature Switzerland AG 2022
T. Montmerle, Y. Zhou, *China and the International Astronomical Union*, Historical & Cultural Astronomy, https://doi.org/10.1007/978-3-031-01787-2_3

Fig. 3.1 Otto Struve, IAU President (1952–1955) (Credit: Kharkiv National University) Kharkiv, (Харків), the Ukrainian name for Kharkov (Russian: Харьков), was Struve's birthplace

as 1946 (Fig. 3.1). The General Secretary was Th. Oosterhoff (1904–1978) (Fig. 3.2), co-administrator of the Leiden Observatory with Jan Oort (more below).

This strategic approach had started even before the Dublin General Assembly. In fact, the necessity of having a balance between the USA and the USSR in the IAU Executive had been implemented as early as the Copenhagen Conference in 1946, by appointing A.A. Mikhailov, from Pulkovo Observatory, near Leningrad, as Vice-President.[3]

The first visible action had been the creation of a Commission (the famous "Commission 38"), to facilitate the "exchange of astronomers" with, as Chair, the British Colonel F.J.M. Stratton (a former IAU General Secretary having been elected for three terms, from 1925 to 1935, before becoming ICSU Secretary General from 1937 to 1952). This happened in 1948 during the first General Assembly after the war, in Zürich, located in a notoriously neutral country: Switzerland. Then came the question of where to hold the next one. The move was incredibly bold: Leningrad, at the invitation of A.A. Mikhailov, in 1951! But (as we would see later about the

---

[3] Blaauw's book, p. 143.

Fig. 3.2 Peter Theodorus Oosterhoof, General Secretary (1952–1958), here in 1961 (Credit: J. van Bilsen, National Archief Nederlands)

admission of Taiwan), the Executive Committee was torn between opposite opinions. Let's look at the brief account given by J.-C. Pecker:[4]

In 1951 the IAU General Assembly was to take place in Leningrad. But the political atmosphere was the one of the Cold War; the majority of the Executive Committee (mainly its President Bertil Lindblad and his General Secretary Bengt Strömgren) decided to annul the General Assembly. A certain number of astronomers protested, among whom of course Minnaert, Schatzman, de Jager,[5] and myself. Our Italian colleagues offered to hold the General Assembly in Rome; but for evident organisational reasons, we had to wait till 1952.

For the record, among the IAU Vice-Presidents at that time two played a particularly important role were the Soviet astronomer Viktor A. Ambartsumian (Vice-President 1948–1955; he would become IAU President at the Berkeley General Assembly in 1961), and Russian-born Otto Stuve (one-term Vice-President

---

[4]Pecker (2019).

[5]Future IAU General Secretary (1970–1973; Chap. 5).

1948–1952, then IAU President, 1952–1955). Ambartsumian's invitation was thus discussed, but clearly (in spite of the support by Pecker and his friends, who were not Executive Committee members!) holding a General Assembly in Stalin's USSR, and just after the Korean War broke out, was certainly premature. So, the Italian Vice-President Giorgio Abetti proposed to organize the next General Assembly in Rome in 1952. This proposal was also doubly symbolic: On the one hand, it was a "family heritage," since Giorgio Abetti was the son of Antonio Abetti, who, after WWI, had been Vice-President in the first IAU Executive Committee and had himself organized the first IAU General Assembly in Rome in 1922, exactly 30 years earlier;[6] and on the other hand, it celebrated the "rebirth" of the IAU after WWII. Then came Dublin in 1955, when the horizon started to clear up on the Soviet side after Stalin's death in 1953. Ambartsumian's invitation was finally accepted, in spite of an ever-present Cold War: *the next General Assembly would at last take place behind the "Iron curtain"* in a country which was the arch-enemy of the USA, and it did take place in spite of the crushing of Hungary (an IAU member since 1947) by Soviet tanks in 1956. It was as an act of immediate reciprocity and good will on the part of the US delegation (headed by Leo Goldberg) to offer, in view of the Moscow General Assembly, to hold the next General Assembly in the USA, but for this Goldberg had to secure the guarantee of the State Department to issue visas to all IAU members, including from Communist countries.

Indeed, the possibility of attending IAU meetings, and in particular General Assemblies, must be guaranteed whatever the country of origin of the participants. In practice, the question (frequent at the time) arose for citizens of countries having no diplomatic representation in the host countries: It was then the responsibility of the host countries to take special steps to facilitate the issuing of visas. This can be done by sending individual official invitations: In the case of the USSR, the inviting body was her Academy of Sciences. As a rule, the procedure did not pose any problem, but there was an obligation: that the participants should be accommodated in designated hotels (so that the KGB could keep an eye on them in various ways: we'll see an example below). For the USA, on the other hand, the invitations are screened by the US State Department, under the authority of the Secretary of State (the US Foreign Minister), who decides who can, or cannot, be admitted in the US territory. This means that the inviting body (the US "National Committee for the IAU," on behalf of the National Academy of Sciences and of the National Research Council) has to negotiate the visa issues at high level with the State Department.[7]

To understand this key point in the case of the USA, one has to go a bit back in time. WWII, as did WWI, had amply shown the importance of science in the conduct of war. The Allied had largely preyed on German scientists to develop a missile program, and of course, there was a fierce competition to develop also a nuclear weapons program, the two benefitting (in the approach of the military at least) from

---

[6]E.g., Blaauw's book (Chap. 4)

[7]This is a continuing policy: For example, it was the case also for the Honolulu General Assembly in 2015.

each other. In this context, the US government created the CIA's "Office of Scientific Intelligence" in 1949, and the "Office of Science Advisor and Special Assistant to the Secretary of State" in 1950, to address the intersection of international science and national security. The Secretaries of State were successively Dean Acheson under the Truman administration, and under the Eisenhower administration John Foster Dulles, succeeded by Christian Hester after Dulles died from cancer on April 22, 1959.

As it turned out, the Office of Science Advisor had progressively lost its importance, until October 4, 1957, when the first artificial satellite, *Sputnik-1*, was launched by the USSR, an event that took the world by surprise. John Foster Dulles then resurrected the Office by appointing in early 1958 as his personal science advisor a prominent scientist, Wallace R. Brode (1900–1974), a chemist and spectroscopist, being, among many other positions, Director of the American Chemical Society, and having been elected to the National Academy of Sciences in 1954.[8] Brode had already played a significant role in Europe during WWII with the "Office of Scientific Research and Development,"[9] which F.D. Roosevelt created on June 28, 1941, to coordinate applications of scientific research for military purposes.[10] Brode was the person interacting with Goldberg,[11] albeit referred to anonymously in IAU documents as "the State Department," from early 1958 (i.e., a few months before the Moscow General Assembly) to the end of the Eisenhower administration, on January 20, 1961 (a few months before the Berkeley General Assembly).

While the organization of the Moscow General Assembly was under way, the American application was progressing. At the next meeting of the IAU Executive Committee (#17: Liège, July 4–6, 1957), then chaired by André Danjon, the French President of the IAU, the past President O. Struve, in his capacity of Advisor, gave some news about the application to host the General Assembly in 1961, on behalf of Goldberg, now chair of the US National Committee in charge of the local organization: The exact venue had not yet been decided, but Goldberg was already in touch with the State Department, and was expected to communicate soon the conditions under which IAU members could participate in the General Assembly. The minutes of the Liège meeting add: *"Members of the IAU living in countries in which there are no USA diplomatic representatives could obtain their visas from another country."* This simple procedure may have been in force at the time (this was before the *Sputnik-1* launch and Brode's appointment), but would simply have been irrelevant in the case of China anyway: The People's Republic was not recognized by the United States, unlike Taiwan which, under the name of "Republic of China" (in continuation of its pre-1949 situation), was the sole representative of China to the

---

[8] McClure (2002).

[9] Sawyer (1974).

[10] Including, as of 1942, the Manhattan Project, but also research on medicine, radars, etc.

[11] Aller (1997). On p.117, it is mentioned that Goldberg (then in his 30s) *"worked successfully on an antisubmarine project at McMath-Hulbert Observatory,"* so, like Brode, contributing himself to some extent to the war effort using his scientific background.

United Nations since its foundation in 1945, and, as one of the victors of WWII, also a member of its Security Council.

Indeed, unrecorded in the Liège Executive Committee meeting minutes is the fact that, on its way to the meeting, Struve went to the American Embassy in Brussels. It is not clear whether he was summoned to go or whether he wanted to check some information, but at any rate Fu&Ye write:[12]

> It was obvious that [Struve] had discussed some IAU political issues with American officials. For example, the soon to be held IAU General Assembly in Moscow and the American invitation process of the next meeting in 1961. The order that Otto Struve received was that he had to take in consideration of the US policy towards Taiwan. In other words, in any international meeting it had to insist that Taiwan was the legitimate representative of China. Thus, the IAU had no choice but invite Taiwan to its General Assemblies, even though it was not a member of the IAU.[13]

Apparently, the pressure from the State Department grew, and the "invitation" turned into a genuine application for membership, perhaps this time under the influence of Brode himself. Quoting again Fu&Ye:

> In March 1958, both Goldberg and Struve extensively discussed with the State Department the idea of holding an international science conference. During their discussions, it was obvious that the State Department asked the President of the IAU to welcome the Taiwanese to become its members. It was distinctively reflected in the letter written by Goldberg that was destined for the State Department on March 24th that year: (I strongly) oppose the idea of abandoning our invitations in exchange of Formosa (Taiwan) becoming a member . . . I believe any attempt at combining the invitation to the Nationalist Chinese and the US's General invitations to the General Assembly would possibly considered by the IAU Executive Committee as political interference. I think this will damage the reputation of US Sciences.

Notwithstanding, a formal application was indeed sent from Taiwan. According to Fu&Ye,

> on April 30th, a month after Goldberg had written to the US State Department, Taiwan applied for membership status. This letter was written by the 'President of the ROC[14] Taiwan's Astronomical Society' to the Executive Committee of the IAU. It roughly mentioned that Taiwan's Academia Sinica had officially started to work with the IAU since early April. And because Academia Sinica is a member of the International Science Council (ICSU),[15] thus 'the ROC Taiwan's Astronomical Society' should naturally have a membership at the IAU, it is their intention to officially apply as well. Soon after, the Executive Committee received letters from Academia Sinica dated May 5 as well. It had proven its official application intent.

---

[12] Its Chap. 5, translated by Yi Zhou (Appendix D).

[13] Blaauw's book (Chap. 9). No reference is given in relation to this episode. The fact that it was not recorded in the Liège Executive Committee meeting's minutes is understandable, since it was not associated with an approach in any form by Taiwan, and explicitly recording a request by the State Department was of course out of the question.

[14] "ROC" stands for "Republic of China." Similarly, "PRC" stands for "People's Republic of China."

[15] Actually, this is not exactly correct: It is true that the Academia Sinica in Tapei was admitted by ICSU in 1958, but that was only in October. More on this point in Chap. 4.

The Academia Sinica, the original Academy of Sciences of the Republic of China, was created in 1928 in Nanking (Nanjing) (Chiang Kai-shek's capital since he took power that year), then moved to Taipei in 1949 following the retreat of Chiang Kai-shek and his government. For the IAU, the adhering organization for China was not the Republic of China, but the Chinese Astronomical Society (CAS), founded in 1922 with its secretariat located in the Beijing Ancient Observatory. In 1932, it moved to the Astronomical Institute, Academia Sinica, in Nanking (the predecessor of the Purple Mountain Observatory, established nearby in 1934), and China adhered to the IAU in 1935.[16] Since the large majority of the members of the CAS, in particular its President Y.C. Chang[17] had decided not to follow Chiang Kai-shek to Taipei, the CAS remained based in Nanking and the membership to the IAU was seamlessly transferred to the People's Republic of China, without any specific conditions: For instance, a Chinese astronomer then working in the UK, Dai Wensai, attended the Stockholm General Assembly in 1938, and similarly Y.C. Chang could attend the Dublin General Assembly in 1955, i.e., well into the first years of existence of the People's Republic.

In other words, for the IAU the distinction between the "Republic of China," admitted 15 years before, and the "People's Republic of China," after 1949, was simply irrelevant.

We will return to this key issue several times below.

3.2   The Moscow General Assembly

Summer 1958 came, and the Moscow General Assembly opened on August 12. It drew a large fraction of the astronomical community (820 participants out of 1100 IAU members), on recently opened air routes[18] (Fig. 3.3). Welcome addresses were given by various speakers, including A.N. Kosygin, Vice Chairman of the Council of Ministers of the USSR, and V.A. Ambartsumian, Academician and President of the Organizing Committee[19] (Fig. 3.4). Spurred by the launch of the first artificial satellites the previous year (*Sputnik-1*, on October 4, and *Sputnik-2*, a month later, carrying the first living being in space, the dog Laika),[20] the General Assembly was a huge success. For the first time, a daily internal "newspaper" ("КОСМОС":

---

[16] At the Paris General Assembly; the same year as the USSR (see Sect. 3.4).

[17] Zhou (2022).

[18] See, e.g., the *Aeroflot* time table of the time: https://aeroflotarchives.com/timetables.html

[19] Transactions Vol. X, p. 5.

[20] To be fair, the first American artificial satellite, Explorer-1 (also named "1958 Alpha"), launched on February 1, 1958, although ten times lighter than *Sputnik-1* (8.4 kg vs. 83 kg) was much more interesting scientifically: Equipped with particle detectors, it had a highly eccentric, high-altitude orbit (538–2550 km), which led to the unexpected discovery of the so-called "Van Allen belts" of energetic particles around the Earth (named after its discoverer James van Allen, from the University of Iowa).

Fig. 3.3 Part of the French delegation to the Moscow General Assembly (Air France flight via Stockholm). Identifications at the time of writing: 1: André Lallemand (Paris), 2: Henri Andrillat (Montpellier; tentative), 3: Audouin Dollfus (Meudon), 4: André Danjon (Paris), IAU President, 5: André Couder (OHP), IAU Vice-President, 6: Renée Cavaggia (Paris), 7: Gaston Fayet (Nice), 8: Mme Chalonge, 9: Daniel Chalonge (IAP), 10: Jean Dufay (Lyon), 11: Jacques Lévy (Paris), 12: Roger Bouigue (Toulouse), 13: Louis Arbey (Paris), 14: Jacques Arsac (Paris). We were unable to identify the other passengers. (Photographer unknown; © Bibliothèque de l'Observatoire de Paris)

Торжественное открытие 10-го съезда Международного Астрономического Союза
Inauguration solennelle de la X-e Assemblée générale de l'Union Astronomique Internationale
Grand Opening of the Xth General Assembly of the International Astronomical Union

Fig. 3.4 Inaugural ceremony of the Moscow General Assembly, in the Column Hall of the House of Trade Unions, on August 12, 1958 (From the "COSMOS" daily newspaper; © IAU)

Сегодня в Москве открывается X съезд Международного Астрономического Союза

C'est aujourd'hui qu'aura lieu à Moscou l'inauguration de la X-me Assemblée Générale de l'U.A.I.

The Xth General Assembly of the International Astronomical Union Opens To-day

КОСМОС

12
1958

АВГУСТ
AUGUST
AOÛT

Газета Организационного комитета
по подготовке X съезда
Международного Астрономического
Союза
Gazette du Comité d'Organisation de la
X-me Assemblée Générale de l'Union
Astronomique Internationale
Newspaper of the Organizing Committee
for the Xth General Assembly of the
International Astronomical Union

NOTRE ASSEMBLEE DOIT ENCORE MIEUX RAPPROCHER LES ASTRONOMES
A. DANJON, Président de l' U. A. I.

Voici enfin cette journée impatiemment attendue où l'Union Astronomique Internationale, répondant à l'invitation de l'Académie des Sciences de l'Union des Républiques Socialistes Soviétiques, ouvre à Moscou sa Xe Assemblée Générale. Des astronomes sont venus nombreux de toutes les parties du monde, pour participer à nos réunions de travail. Mais quelque chose de plus les attirait à Moscou: la perspective de nouer avec leurs collègues de l'Union Soviétique des liens de confraternité scientifique plus étroits. Le rôle de l'Union Astronomique Internationale n'est pas seulement de constater les progrès récents de l'astronomie et d'élaborer des plans de recherches, mais encore de rapprocher ceux qui, à travers le monde, s'occupent des mêmes sujets. A cet égard, cette Xe Assemblée Générale qui réalise la conjonction des astronomes de l'Est et de l'Ouest, aura sans aucun doute les plus heureuses conséquences, pour le progrès de l'Astronomie.

НАШ СЪЕЗД ДОЛЖЕН ЕЩЕ БОЛЬШЕ СБЛИЗИТЬ АСТРОНОМОВ
Президент МАС проф. А. ДАНЖОН

НАУКЕ НУЖЕН МИР
В. АМБАРЦУМЯН,
председатель Организационного комитета по проведению X съезда МАС

SCIENCE NEEDS PEACE

In the three years that have passed since the General Assembly of the IAU was held in Dublin astronomy made a great stride ahead. It suffices to see what wonderful means of research appeared in this period of time. A gigantic radio-telescope in Manchester, a few very large new optical telescopes and finally the artificial Earth satellites that have

LA SCIENCE AU SERVICE DE LA PAIX
V. Ambartsoumian
President du Comité d'organisation de l'U.A.I.

V. AMBARZUMJAN, Chairman of the Organizing Committee of the IAU

opened the epoch for the mastery of the Cosmos by man are now of aid to us in the solution of astronomical problems.

I presume that in the following

Fig. 3.5   Front page of the "КОСМОС" ("COSMOS") daily newspaper (© IAU). This was the first time such a daily newspaper was published. Every General Assembly had its own thereafter (albeit not in three languages). *Left:* André Danjon; *right:* Viktor Ambartsumian

"*Cosmos*") was issued, with articles in three languages (Russian, English and French, the last two being the official languages of the IAU) (Fig. 3.5). The American invitation to host the XIth General Assembly in 1961 was officially accepted, even though the venue was still undecided (East Coast or West Coast?).

But behind the scenes, at the IAU executive level, and within the Chinese delegation, dramatic events were unfolding on how to handle Taiwan's application.

In the IAU Archives in Paris, we discovered an unpublished "personal account" written (by hand!) in 1976 at the request of IAU General Secretary P. Wayman (who wanted a summary of the history of the "China problem" after the Grenoble General

Assembly—see Chap. 7) by Donald Sadler, IAU incoming General Secretary at the Moscow General Assembly:[21,22]

> The President [A. Danjon] convened, at very short notice, *a meeting of certain members of the present and future Executive Committees, excluding nationals of U.S.S.R. and other 'eastern' countries.*[23] The meeting was held in a private room at the Hotel Ukrainia and *(in spite of a note from myself requesting restraint -I had been briefed by the Foreign Office!)* was extremely frank. The President announced that he had received, by hand, a communication from a high authority in U.S.A. [I never saw this and do not know whether it was actually addressed from the State Department, or was passed by the Academy of Sciences] to the effect that the U.S. government would withdraw all support for the promised invitation [for the Union to hold its G.A. XI in California in 1961] if the IAU did not admit the Republic of China to membership during G.A. X in Moscow. *Leo Goldberg (nominated for the future Executive Committee), who I think may have known about this communication, made it clear that, without government support (especially as regards visas) it would be impossible to hold G.A. XI in USA.* During the meeting, the opinion grew that the original application from Taiwan must have been originated by the U.S. State Department. *There could be no doubt that the Union was being blackmailed in the most blatant manner. Danjon, quite justifiably, was furious and would, I think, have been prepared to reject the application outright.*
>
> However, it was agreed that
>
> 1. Consideration of Taiwan's application should be deferred until the information requested had been received;
> 2. *The communication from U.S.A. should be ignored;*
> 3. Leo Goldberg should be authorized to 'leak' this information to the appropriate authorities in USA (an unpleasant task!).
>
> I do not know whether the U.S.S.R. members of the Executive Committee were aware of this particular piece of blackmail, *though it could reasonably be assumed that the discussions in the private room at the Ukrainia Hotel could have been monitored.* Certainly, no reference was made to it at subsequent meetings of the Executive Committee; but the General policy of the U.S. State Department was well known.

[In this quotation, as well as in all quotations that will follow in the text, the italics are mine, unless stipulated otherwise.]

Then, in parallel, developments were unfolding on the Chinese side, which (to my knowledge) were also untold until the publication of the book by Fu&Ye. The full text is given in Appendix D, but it is worth quoting here *verbatim* large excerpts from their story:

> The Chinese Mainland delegation arrived in Moscow on August 9. Soon after their arrival in the Soviet capital, they received one good news and one bad news from the then vice-President of the IAU, the Russian astronomer Boris Kukarkin (1909–1977), as well as Viktor Ambartsumian. The bad news was that there was the application from Taiwan, and

---

[21] D.H. Sadler (1908–1987). See Sect. 3.4.

[22] The whole document was typed by TM. Here and there some words were illegible. Excerpts of the original, handwritten document are given in Appendix A.

[23] The "excluded members" were Boris Kukarkin (USSR, outgoing and incoming Vice-President), Eugeniusz Rybka (Poland, outgoing Vice-President), and Bohumil Šternberk (Czechoslovakia, incoming Vice President). The italics are my emphasis, see below.

the good news was that the Socialist Bloc had recommended the Mainland astronomer Cheng Maolan to become the vice-President of the IAU.[24] He was serving as the logistic section chief of the Beijing Observatory at the time.

Under the Cold War context in the early 1950s, the IAU Executive Committee's members were diametrically recommended by the two embattled blocs. Each side could name 2–3 candidates, if no opposition was risen by the other side, then his/her nomination would be subsequently approved.[25] Cheng Maolan was thus an excellent candidate to represent the Mainland because he just recently returned from France in 1957. And he started to study and work in France as early as in 1925. (. . .) Since Cheng had lived in France for 32 years, he had accumulated a strong friendship and network with French astronomers. This special relationship between them was sought after to be a useful advantage when his nomination was considered. This was the reason that the Socialist Bloc wanted it to nominate him.

After learning about these two major pieces of information, they were then treated seriously by the Chinese delegation at the General Assembly. *The deputy Director of the Purple Mountain Observatory Zhu Renjun, who was also in charge of the Party affairs of the delegation quickly reported this information to the Chinese ambassador Liao Xiao in Moscow. Zhu Renjun also served as the deputy Secretary of the communist party branch at the Purple Mountain Observatory. Soon after, the embassy received orders from the Chinese Ministry of Foreign Affairs in Beijing. Beijing asked them to firmly oppose the invitation of Taiwan and to give up the nomination to the vice-presidency of the IAU.* The first order came as expected, however the second one was rather impossible to understand. It caught all of the Chinese delegates by surprise. They thought that having a Mainland scientist at the table of the Executive Committee would be beneficial to solve the Taiwan problem. As one of the delegates, Wang Shouguan[26] recalled, all members from the then Mainland delegation were largely disappointed, so was Kukarkin. The head of the delegation, Zhang Yuzhe (Fig. 3.6) who normally was rather a reserved person, even he expressed strong reactions. After more than half of century, Wang Shouguan thought maybe in retrospect, this was the right thing to do. *As politics was deeply involved and played a huge role in academic organizations, the two opposing blocs were already right against each other. Under these complex circumstances, Cheng Maolan who just came back to China at the time would not have been able to handle by himself such a sensitive issue as Taiwan. Besides, there were many other risks as well, it was a better idea to give up this nomination.*

After having received orders from the Chinese embassy, the Chinese delegations made various preparations for the possible scenarios prior to the General Assembly. In the worst-

---

[24]This statement triggered a detailed investigation about Cheng Maolan (Tcheng Mao-lin, 程茂兰; 1905–1979), a Chinese astronomer who had spent 32 years in France before returning to the motherland in 1957 (see his biography by Montmerle et al. 2022) [MZG22].

[25]I have been unable to find a reference to this surprising statement. At least, there is no mention of such a "political recommendation" in the minutes of the Executive Committee meetings from 1950 (#12, Stockholm) to 1957 (#17, Liège), and it would clearly be against the rules of the IAU. At that time, future members were nominated by the Executive Committee only (after appropriate consultations), certainly not by countries or "blocs." In practice, during this period, there were never more than two Vice-Presidents (out of six) from Eastern countries, but always, indeed, one Soviet Executive Committee member (not necessarily from Russia; see Sect. 3.1). In my opinion, it is unclear whether the USSR would have considered China as equivalent to an "Eastern country" anyway, especially at a time when the relations between the two countries had started to deteriorate. See EC minutes #18, Moscow: item 4 (https://www.iau.org/static/administration/ecminutes/ec1 9minutes.pdf).

[26]See his personal recollections (Shouguan 2022).

~~~~~~

ASTRONOMY IN NEW CHINA

It is well known that the Chinese people began to study celestial phenomena from early antiquity. Length of the year was found to be 365 1/4 days and eclipses were predicted four thousand years ago.

Of course, we could not be satisfied with our past glories. Since the revolution in 1949, the Purple Mountain Observatory in Nanking has made a large number of photographic observations of comets and asteroids. The Zi-Ka-Wei Observatory in Shanghai takes charge of time service. It is equipped with a complete set of quartz-clocks, photo-electric transit instruments, impersonal astrolabe à prisme, etc. Seven times each day it broadcasts time signals of high precision. Zo-Se Observatory near Shanghai participates in the cooperative work of Faint Star Catalogue. Near Tientsin, along the parallel circle 39°8′, a new latitude station, equipped with a 180-mm zenith telescope has been established. In the suburb of Peking, we recently set up a chromospheric telescope, which has begun to make routine observations of the sun in monochromatic light. There we have also just completed the installation of three radio telescopes, for solar radiation of wavelengths 2 cm, 3,2 cm and 3 meters respectively. The results of our investigations and observations in astronomy are published in the journal "Acta Astronomica Sinica". Six issues of this journal have appeared up to the present moment.

A more ambitious project is the establishment of Peking Observatory. Its equipment includes a reflecting telescope of 2-meter aperture, a 600 mm/900 mm Schmidt camera, and a twin astrograph of 400-mm aperture. These instruments are already under construction. The observatory buildings as well as the installation of these telescopes will be completed by the year 1962. The principal work of this observatory will be stellar spectroscopy and photometry.

A high altitude observatory on the Tibetan Plateau for solar observations and an observatory in the South for general purposes equipped with large telescopes are among the plans considered. But they will come a little later. For the time being, we shall concentrate our efforts on the projects that have already been under way.

In developing our astronomical work, Soviet astronomers have rendered invaluable help and we welcome suggestions and help from our friends all over the world. At the same time, we are willing to do our share in the family of astronomers of the world and participate in their cooperative undertakings.

CHANG YU-CHE,
director of the Purple Mountain Observatory

18	Août	COSMOS	3
18	August	COSMOS	3
18	августа	КОСМОС	3

Fig. 3.6 Y.C. Chang ("Chang Yu-Che" here; now Zhang Yuzhe, 张钰哲) (See the "*Note on Chinese names*" ending Chap. 2), head of the Chinese delegation at the Moscow General Assembly, describing the projects of the "New China" in astronomy in the "KOCMOC" daily newspaper (issue of August 18; © IAU)

case scenario, they were ready to completely withdraw from this meeting, if the Taiwan issue was ever raised. However, this was not the case.

On August 19th, the Executive Committee held a meeting, it made the decision that the Taiwan membership would be pushed back to be discussed in the next Executive Committee meeting under the direction of the newly elected President Jan H. Oort. They also required more information about Taiwan's astronomical activities.

The letter further explained that it was a misunderstanding to think the Taiwan membership issue would be resolved in the Moscow General Assembly. In addition, Danjon also raised some questions about the level of astronomical activities in Taiwan. For example, he pointed out that the IAU did not know where Taiwan's observatory was located. However, he promised that the Executive Committee of the IAU would reconsider Taiwan's application in 1959. It was clear Danjon did not mention whether this next meeting would decide on Taiwan's membership, nor did he outright reject this possibility.

So, behind the scenes of a happy General Assembly, the Cold War reality was that, on the one hand the State Department applied pressure ("blackmail," even says

Sadler) on the Executive Committee to admit Taiwan immediately, in exchange of a promise of not applying visa restrictions on Chinese astronomers wishing to attend the following General Assembly in the USA, and on the other hand, the Chinese delegation receiving orders from the Chinese embassy in Moscow to withdraw from the General Assembly (thus embarrassing the USSR, the host country) if the Taiwan proposal came to be discussed. (Here this side of the information was obviously channeled to the Executive Committee via the Soviet side, Kukarkin and/or Ambartsumian.) Either way, in Moscow the Executive Committee was trapped at short notice between two opposing political forces.

3.3 Cold War Context: The 1960 Winter Olympics in California

For the time being at least, the recommendations secretly agreed in the Hotel Ukrainia by the "non-Eastern" Executive Committee members would be followed, therefore resisting the American interference (*"The communication from U.S.A. should be ignored"*.).

A reason allowing this "resistance" to be a successful option may have been that this interference had actually diminished in intensity, because of a seemingly unrelated event: the forthcoming Winter Olympics, due to take place on February 18–28, 1960, in Squaw Valley, a famous ski resort just a few miles away from Lake Tahoe in California (and about 300 km by road from Berkeley, where the next IAU General Assembly would take place the following year).

This surprising connection is explained by Lawrence Aller (who had worked on the Manhattan Project and had chaired the Astronomy Department at UCLA from 1963 to 1968). In his Obituary of Leo Goldberg, he wrote:[27]

> The cold-war rivalries in the United States and the Soviet Union flourished in the 1950s. The Soviet Union invited the IAU to convene in Moscow in 1958 and proclaimed that astronomers from all member countries would be welcome. To save face, the United States would have to host the meetings in 1961 under the same guarantees that the Soviet Union had given. Astronomers from all member countries would be welcome. In particular, since the People's Republic of China (PRC) was an IAU member, its astronomers would be allowed to attend, but Taiwan was not a member at that time. *PRC athletes were to attend the 1960 winter Olympic games in the United States.* The Department of State headed by Secretary Dulles observed that this action was ad hoc and did not imply that mere scientists could expect such favors. Goldberg contacted his representative in the U.S. Congress, George Meader, a conservative and fair-minded Republican, who presented the case to Dulles, who referred it to his science advisor Wallace Brode. Brode promptly demanded that Taiwan be invited to the IAU.

This link between the IAU and the Olympic Games may seem artificial, but it is not (and in fact the above lines from Aller suggest that the visa issue of the 1960

---

[27] Aller (1997), p. 129.

Winter Olympics was a reason for bringing the issue of Taiwan to the IAU's attention). Clearly, from the action of Goldberg and of the US National Academy of Sciences (NAS), which strongly supported him,[28] the issue of the "US prestige" for the astronomers eventually played a positive role: even though the impact on the public worldwide would of course be far greater in the case of the Olympics, the impact on the world scientists (not only the astronomers) of not inviting the PRC would also be clearly damaging in the case of the IAU.

The comparison may be pushed even further and the background is worth describing. Indeed, the Olympic Games and the IAU share the same "universal" values (as does ICSU, which issued a Resolution in 1958 defining the first "basic policy of political non-discrimination" between scientists of any country, whatever its political regime[29]). The similarities between the "China problem," as faced by the IOC (International Olympic Committee) and the IAU during the preparation of their respective events, are striking.

In his book *The Politics of the Olympic Games*,[30] the historian Richard Espy writes (pp. 61 et sq.):

> Competition between the two countries [USSR and USA] and their respective blocs, East and West, remained intense. *There were still opposing interests on various issues, such as Germany, China, and Korea.* In the Olympics, several of these issues were exhibited. Perhaps the most controversial issue along this line was the question of Chinese participation.
>
> Developments regarding the PRC in the Olympics over the period 1956 to 1960 closely followed developments in the political arena. The Winter Games of 1960 were scheduled to take place at Squaw Valley, California, and there was some concern in the IOC over whether the United States would curb its restrictions and admit the PRC, a country the United States did not recognize. This problem was not limited to the PRC but extended to Communist countries in General. In 1957, Brundage[31] warned the United States that if it refused any properly IOC-recognized country the right of participation at Squaw Valley, the IOC would be forced to revoke the Games from the United States and award them elsewhere.

After some arm-twisting between Brundage and the State Department,

> Participation in the 1960 Olympic Games seemed possible for all Communist countries recognized by the IOC but with which the United States did not have diplomatic relations.
>
> (. . .) In 1957 there was every reason to believe that the PRC intended to participate in 1960 unless the Taiwan issue intensified. Despite absence from the Melbourne Games [in 1956], the PRC had participated in numerous sport events. The climate in the People's Republic was favorable with Mao's decree to 'Let a Hundred Flowers Bloom'.[32]

---

[28] Goldberg was elected to the NAS in 1958. This may have helped Brode to be more receptive to the position of the Academy.

[29] See Chap. 4.

[30] Espy (1979). The book covers the period from the XIIIth Olympiad (1944–1948) to the 1976 Olympic Games, so unfortunately not the period in which China re-joined the IOC (Chap. 7). In all, China did not participate in the Olympics from 1956 until 1984.

[31] Avery Brundage (1887–1975), an American, was President of the IOC from 1952 to 1972.

[32] A 1-year period (1956–1957), in which freedom of expression and criticism was encouraged by Mao.

In others words, Taiwan and China had been somehow co-existing and competing on sports fields for quite some time.

This peaceful situation had been the result of a decision taken by the IOC in 1954, at its annual session in Athens. Espy explains:[33]

> In a close vote, 23 to 21, the IOC chose to recognize both committees [of Mainland China and Taiwan (Formosa)]. In this way, they followed the political developments of the day. *Their ruling, for all intents and purposes, recognized two separate states: Peking-China and Formosa-China.* To that extent they did not sacrifice their rules, for the rules stated that there could be only one committee per country. *By recognizing two committees they recognized two countries.* Entry in the realm of politics was contrary to the Olympic principles and to the Olympic charter, so if it took a neutral stand, as the IOC did, the ruling was in conformity with its regulations and principles'. (. . .).
>
> Developments in the IOC regarding the Chinese issue proved otherwise. (. . .) The recognition of China was an expedient. (. . .) The best solution was to recognize both [China and Taiwan] as the world did in a fashion and hope that the situation would settle. The situation would not, however, and did not—to the continual frustration of the IOC.

This is indeed what happened in 1958 at the IOC, when the situation changed dramatically and events occurred exactly in concordance with the Moscow General Assembly. Probably unbeknownst to the astronomers, these events foreshadowed in a striking manner the consequences of the admission of Taiwan by the IAU:[34]

> by 1957 and 1958 the PRC authorities had retrenched and were beginning the 'Great Leap Forward'. The retrenchment became immediately obvious in Olympic circles. After a relatively quiescent period following the Melbourne Games, the PRC began a barrage of demands for the expulsion of the Taiwan committee from the Olympic movement. Throughout that year IOC President Avery Brundage and Tung Shou-yi, the PRC member on the IOC, corresponded on the issue. *Brundage emphasized that two committees existed, that political issues were not the province of the IOC,* and that Tung violated his obligations as an IOC member by constantly raising political issues in the meeting and correspondence. Tung replied with demands for Taiwan ouster and stated that *it was not he, but Brundage, who was introducing politics by the continued insistence on two committees.* Finally, in a letter dated August 19, 1958, (. . .)

[Tung told Brundage]:

> 'I will no longer cooperate with you or have any connection with the IOC while it is under your domination'. *With that, the PRC withdrew from the IOC and, at the same time, from all the international federations with which they were affiliated.*

Back to the IAU, as a result of all these Cold War twists and turns, when the Executive Committee meeting (#18) took place in Moscow (in two sessions: August 11 and August 19, 1958, the very same day China pulled out of the IOC), the agenda item 7 (out of 30) simply mentioned: "Request from Taiwan for IAU membership." The minutes consisted of only a few lines, gave no details, made no reference to a pressure by the US State Department (as could be expected), but only mentioned that *"very little information had been given concerning the National Committee of*

---

[33]Espy (1979), p. 45.

[34]Espy (1979), pp. 62–63.

*Astronomy in Taiwan and about astronomical activity in this country, the Executive
Committee decided to postpone a decision in this matter until the next meeting of the
new Executive Committee."* The minutes added a short paragraph: *"The President of
the IAU submitted to the meeting a draft reply to the Academia Sinica in Taiwan. As
the President asked for a vote on this letter,*[35] *all the members were in favour of this
reply, with the exception of Prof. Kukarkin who abstain from voting."*[36]

By doing so, as mentioned in Blaauw's book,[37] Danjon, quoted many years later
by the Belgian IAU President Pol Swings (1964–1967), allegedly said *"exercising
the President's role was like an aviator always on high alert so that he won't drive
the plane into a cliff."* That was good for him, since he was outgoing President
(while still being an advisor), but less so for his successor, the incoming President
Jan Oort, and the new Executive Committee.

3.4   The New Executive Committee: A Highly Experienced Leadership

It is perhaps appropriate to start by a brief presentation of the members of this
Executive Committee (which is defined by the IAU Statutes as a group of ten
astronomers: the President and the General Secretary, six Vice-Presidents and two
non-voting Advisors, the past President and the past General Secretary) that would
be in charge of the IAU affairs—and especially of the highly contentious China–
Taiwan case—for the next triennium, from the Moscow General Assembly to the
Berkeley General Assembly.

Its first meeting after the Moscow General Assembly (#19) took place on
September 7–9, 1959, in the library of the beautiful castle of Herstmonceux
(in East Sussex, England, near Brighton), where the Royal Observatory, originally
situated at Greenwich, had been recently relocated (Fig. 3.7). The reason was that the
castle was home to H.M. Nautical Almanac Office (HMNAO), of which Donald
Sadler, the General Secretary, was Superintendent.[38]

Donald Harry Sadler was born in 1908 in Yorkshire, England. He developed an
interest in celestial mechanics while studying in Cambridge, from which he graduated in 1929. He became HMNAO's Superintendent in 1937, until his retirement in
1972. He acted as Secretary of the Royal Astronomical Society from 1939 to 1947,
running its activities throughout the war (Fig. 3.8). Directly relevant for us is the
appreciation given by G.A. Wilkins, Sadler's successor, in his Obituary (Wilkins
1991): *"Sadler felt very strongly that the activities of all such organizations, whether*

---

[35] This letter seems to have been lost.

[36] The implication is that the Polish Vice-President, E. Rybka, also voted positively.

[37] Blaauw (1994), p. 193.

[38] H.M. stands for "Her (or His) Majesty." This Office was established in 1832 under King William
IV. See Wilkins (2008).

Fig. 3.7 Herstmonceux Castle, view from the southeast. It was built the same year (1441) as King's College, Cambridge, under King Henry VI. (License: Creative Commons Attribution-Share Alike 4.0 International: https://commons.wikimedia.org/wiki/File:Herstmonceux_Castle_with_moat.jpg)

Fig. 3.8 Donald Sadler, as President of the RAS (1967–1969) (© Royal Astronomical Society)

a local sports club or an international scientific union [meaning here the IAU], *should be conducted in accordance with an appropriate set of rules. He not only took time to draft new or improved rules, but he would draw attention to situations where he felt that the rules were not followed correctly; to some this appeared pedantic, but he was only too well aware of the difficulties than can arise when such rules are broken."*

Remarkably, Sadler would be the second longest serving IAU General Secretary: 12 years, from 1955 to 1967 including his years as Assistant General Secretary before the Moscow General Assembly (1955–1958), and as Advisor (1964–1967) after the XIIth General Assembly in Hamburg. The longest serving GS was Jan Oort (17 years, if including the war years, see below).

Incidentally, Sadler, in his *Personal history of H.M. Nautical Almanac Office* (edited by Wilkins, 2008; p. 115), recalls that he had been invited in 1954 to attend the re-opening of Pulkovo Observatory, which had been completely destroyed by the Wehrmacht during the siege of Leningrad: He mentions that he met Th. Oosterhoff (who would be elected IAU General Secretary at the Dublin General Assembly the following year),[39] and J. Oort (then in between two long terms as General Secretary and as President, see below), and many Soviet friends, including "the Kukarkins"— Boris himself being now second-term Vice-President. During this event, he may also have come across Y.C. Chang, the President of the Chinese Astronomical Society, who would be "on the other side" when the Taiwan issue was put on the table.[40] Small world . . .

The other major figure of this Executive Committee was Jan Oort himself, its President (Fig. 3.9). When he took office in Moscow, he was already widely known, with an exceptional stature, both as an astronomer having made impressive discoveries, and for having a long experience in IAU affairs.

Jan Hendrik Oort was born on April 28, 1900, in Franeker, a small village located about 50 km West of Groningen, in the northern Dutch province of Friesland. His family moved to Oegstgeest, a suburb of Leiden, near The Hague, when he was three years old. In 1917, he went to study physics at the University of Groningen, where a famous astronomer, Jacobus Cornelius Kapteyn, was teaching astronomy. Kapteyn's pioneering work on bright and dark regions of the Milky Way (competing in part with the ongoing *Carte du Ciel* project)[41] aroused the interest of the young Oort, who would make the study of the Milky Way one of his main themes of research, leading to major discoveries. Oort undertook graduate studies at Yale University in 1922, making his first astronomical observations as an Assistant to Franck Schlesinger, the Director of its Observatory and future IAU President (1932–1935). He returned to Leiden in 1924 at the request of Willem de Sitter, the Director of the Leiden Observatory, who would become IAU President the following year and was famous for his work with Albert Einstein on Relativity and

---

[39] Section 3.1.

[40] See Sects. 2.4, 3.7, and Zhou (2022).

[41] See Chinnici (2022).

Fig. 3.9 Jan Oort in 1956 (Credit: Leiden Observatory, with permission). Jan Oort is here seated behind the photographic refractor of the Old Observatory in Leiden. The picture was presumably taken by Herman Kleibrink, when measurements of the optical polarization of the Crab Nebula were taken by J. Oort and Th. Walraven. It was a rare occasion for Oort to sit behind a telescope at the time. (Information kindly provided by Drs. Alexandra Schouten-Voskamp, and collected by Frank Israel, Emeritus Professor at Leiden University.)

Cosmology. Oort then went back to Groningen to defend his PhD thesis (1926), before returning for good in Leiden, where he was eventually appointed *Professor Extraordinarius*[42] in 1935.

At that time, he already had a high scientific reputation, after major achievements such as having established (in 1927) that the Milky Way was in fact a gigantic stellar system rotating about a center (in the constellation Sagittarius) located 20,000 light-years from the Earth.[43] Another major discovery (1950) was the recognition that comets must originate from a vastly extended "cloud" (now known as "Oort's cloud") of primitive asteroids formed at the birth of the solar system, extending up to 3 light-years from the Earth.[44] At the end of WWII, he became also well known for having initiated the discovery by H. van de Hulst of the spiral-arm structure of the Milky Way by use of the 21-cm line of neutral hydrogen, thus giving birth to

---

[42] "Professor without chair," a lifetime position borrowed from the German system.

[43] Using data available at the time. This distance has now been revised upwards to 26,000 light-years.

[44] Equivalent to 2000 times the size of the solar system.

radioastronomy in the Netherlands,[45] and making the country a leader in the study of this new window to the universe.[46]

In Leiden, Oort had a close exposure to the IAU: The Netherlands had adhered to the IAU at its first General Assembly in Rome in 1922,[47] de Sitter being elected to its Executive Committee as Vice-President, then as its President in 1925, and the University had been the venue of the IIIrd General Assembly in 1928. It is then perhaps not too surprising that Oort would be proposed in due time for a high office in the IAU, so to speak as the successor of de Sitter (who passed away in 1934). The opportunity came in 1935, as soon as he had obtained his high-visibility professorship: He was too young to be already Vice-President, not to speak of President, so he was elected General Secretary at the Vth General Assembly in Paris. At 35, he would be the youngest IAU General Secretary ever: His two predecessors, Alfred W. Fowler and Frederick J.M. Stratton, both British and having served respectively two and three terms, were 51 and 44 when they were elected.[48]

The Paris General Assembly was an important one, with the adherence of the USSR (via its Academy of Sciences) and China[49] (via its National Committee for Astronomy in Nanking). It is thus somewhat ironic that Oort would witness the adherence of China to the IAU when he took office as General Secretary, to see her leave when in charge as President 25 years later!

Subsequently, as recalled in the Introduction, Oort served the IAU for a long time almost without interruption: While his expected second term (1938–1941) would be shortened by the war (the American Walter S. Adams replacing him during the German occupation of the Netherlands, from 1940 to 1944), it resumed during the 1944–1948 period (until the VIIth General Assembly in Zürich), followed by a term as Advisor during the following triennium, which in practice lasted four years (1948 until 1952, the VIIIth General Assembly in Rome[50]). Oort then declined to succeed the Swede Bertil Lindblad as IAU President, a task which Otto Struve accepted (see above), and took two triennia off (1952–1955, and 1955–1958) with no IAU responsibilities (but actually he was in parallel the Director of the Leiden Observatory, from 1945 to 1970, and was working on comets). Interestingly, his successor as General Secretary in 1952 was none other than Th. Oosterhoff, his co-administrator, so the job stayed in Leiden! Eventually, Oort accepted the "supreme task" and was elected President in Moscow. All in all, he had accomplished (excluding the war years, but including when he was Advisor) 13 years as General Secretary of the IAU.

---

[45] The antennas used were so-called "Würzburg"-type coastal radars left behind by the Germans.

[46] Several obituaries were published after his death in 1992 (e.g., Wilford, 1992; Blaauw & Schmidt, 1993; van de Hulst, 1994).

[47] E.g., Montmerle (2019).

[48] Since then, no General Secretary was elected at a younger age. The two youngest following Oort were the French J.-C. Pecker (1964–1967) and the Dane R.M. West (1982–1985) who were both 38 when elected (at the time as Assistant General Secretaries).

[49] For some reason, contrary to the USSR, the adherence of China is not recorded in the Transactions Vol. V (p. 5), but does appear as having taken place in 1935 in the Transactions Vol. VI (Stockholm, 1938, p. 5).

[50] Replacing the initially planned 1951 General Assembly in Leningrad; see Sect. 3.1.

Therefore, there can be no doubt that the new Executive Committee would be under the leadership of two exceptionally competent personalities, fully qualified to take responsibility for coping with an exceptional chapter of the history of the IAU: the admission of Taiwan and the subsequent withdrawal of China.

In addition to the IAU Officers (Oort as President and Sadler as General Secretary), the other Executive Committee members were as follows: the Vice-Presidents Leo Goldberg (first term, the organizer of the Berkeley General Assembly three years later, and who would play an essential role in the US–China relations, especially during his term as IAU President, from the 1973 General Assembly in Sydney to the 1976 General Assembly in Grenoble, see Sect. 5.3), Otto Heckmann (Germany; second term), Boris Kukarkin, whom we have already mentioned (USSR; second term), Robert Petrie (Canada; first term), Bohumil Šternbek (Czechoslovakia; first term), and Richard Stoy (South Africa; first term).[51] Another Soviet astronomer, (Ms) P.G. Kulikovksy, attended the meeting as an interpreter for B. Kukarkin.

3.5 Herstmonceux Castle: The 19th Executive Committee Meeting

When the new Executive committee met in Herstmonceux, on September 7–9, 1959, unfortunately in the absence of the Advisors A. Danjon and Th. Oosterhoff, and of Vice-President R. Petrie, due to illness, the main item of the meeting (#4) was of course Taiwan's application for membership.

The minutes of this meeting are readily available[52] and have been largely commented in Blaauw's book. It is therefore perhaps more interesting to quote here the backstage summary of these minutes that D. Sadler gave in his 1976 "personal account," because it contains indeed some personal reflections; I will comment on some of them (in italics) afterward and come back to some points of the minutes.

D. Sadler wrote:

Correspondence with the Academia Sinica and the National Committee of Astronomy in Taiwan had revealed a very small amount of astronomical activity (at about amateur level) and some real interest in the history of astronomy in Taiwan. *Even this was more than that in at least one of the National Members (Venezuela) and clearly did not constitute an adequate reason for rejection of the application for membership.* My interpretation of Statute No. 3 was that there was no valid reason, or cause, for rejection, bearing in mind that:

(a) *Taiwan was an adhering member of ICSU;*
(b) It fulfilled all the technical conditions, including having a National Committee of Astronomy, recognized by the Academia Sinica;
(c) There was some independent astronomical activity;

---

[51]Dates: Goldberg (1913–1987), Heckmann (1901–1983), Kukarkin (1909–1977), Petrie (1906–1966), Šternbek (1897–1983), Stoy (1910–1994).

[52]https://www.iau.org/static/administration/ecminutes/ec19minutes.pdf

(d) It could not be excluded, on the grounds that it was a 'country', within the terms of the Statute, *especially as no other 'country' had* de facto *control of astronomical activity within Taiwan.*

The question was discussed at considerable length, with the full knowledge that The People's Republic of China (under the name of China), which had been a National Member of the Union since 1935, would immediately withdraw if Taiwan was admitted; Oort, Goldberg, Stoy and myself also knew that there was a high probability that the invitation to hold G.A. XI in Berkeley in 1961, already made and adopted, would be withdrawn if Taiwan's application was rejected. Most of the discussion centered on (d) above, but it was (a) that was unanswerable: the IAU adhered to ICSU and its statutes, especially in respect to National Members, were linked with those of ICSU. *It would have created much confusion if the IAU had interpreted 'country' in a manner contrary to that which ICSU had already adopted.*

Nevertheless, I felt strongly opposed to the blatant use of blackmail by the U.S. State Department, and tried to draft a resolution that might enable China to maintain its adherence to the Union.

In other words, the Executive Committee was clearly aware that it would have to navigate between *three* opposing hurdles to reach a decision on the admission of Taiwan to the IAU: a scientific one (the "level" and the "independence" of its astronomical activity); and two conflicting political ones: On the one hand, the pressure from the US State department, although, as we have seen and by all accounts this pressure had "gradually mellowed"[53] even before the Moscow General Assembly, a year before (yet Sadler was still feeling uneasy about it), and on the other hand, the mounting pressure from China herself, threatening to withdraw from the IAU should Taiwan be admitted.

3.6 The IAU Statutes in 1958

The only possible way out was a "legal" one, i.e., to use the IAU Statutes (its "Constitution") as a basis to try and reconcile opposing forces. The relevant Sections are I ("Objects of the Union"), and II ("Adherence of Countries"). Here they are, in their 1958 version (Fig. 3.10), as they appeared in the Transactions published after the Moscow General Assembly:[54]

The first two articles of the first section (I. *"Objects of the Union"*) are straight-forward; article 2 (*"The Union adheres to the International Council of Scientific Unions"*) is the one relevant to our discussion, as follows.

1. On the one hand, China (the People's Republic) was considered by ICSU as a "legitimate" member, in continuation of its adherence in 1937 (see Chap. 4), and Taiwan had been admitted separately. On the other hand, the IAU had adhered to

---

[53] Blaauw' exact sentence in his book (p. 192) mentions a "gradual mellowing of the position of the State Department"—see above.

[54] Transactions Vol. X, p. 731.

INTERNATIONAL ASTRONOMICAL UNION

STATUTES

I. OBJECTS OF THE UNION

1. The purpose of the Union is:
   (i) to facilitate the relations between astronomers of different countries where international co-operation is necessary or useful;
   (ii) to promote the study of astronomy in all its departments.
2. The Union adheres to the International Council of Scientific Unions.

II. ADHERENCE OF COUNTRIES

3. A country which adheres to the International Council of Scientific Unions may adhere to the Union (*a*) by the organization through which it adheres to the International Council, or (*b*) by a National Committee of Astronomy formed by or recognized by such organization.
   A country not adhering to the International Council may adhere to the Union by a National Committee of Astronomy recognized by the Executive Committee of the Union.
   The term 'country' is to be understood as including Dominions, diplomatic Protectorates, and any dependency in which independent astronomical activity has been developed.

Fig. 3.10   The first two articles of the IAU Statutes in 1958 (Transactions Vol. X, p. 731)

ICSU since its creation as the International Research Council in 1919. Therefore, the IAU Statutes, especially with respect to National Members, were linked by construction to those of ICSU.[55] Sadler adds: *"It would have created much confusion if the IAU had interpreted 'country' in a manner contrary to that which ICSU had already adopted."* Surprisingly, this link between the admission of Taiwan and its membership to ICSU is not mentioned in the Executive Committee minutes. Perhaps the reason is that this argument, while strong in absolute terms with respect to the ICSU community as a whole, was weak for the IAU itself: as recalled above, Taiwan had just become a member of ICSU, whereas the People's Republic was de facto already an IAU "country" via its Astronomical Society, and Taiwan wanted to be another "country" via its own Astronomical Society. This is precisely the issue discussed in the Statutes, Section II.

2. Section II ("Adherence of Countries") has only one article (#3). This article however contains sentences that are less straightforward, and their interpretation has been at the core of China and the IAU having strongly diverging views over the next 20 years.

---

[55]These Statutes had been adopted at the IVth General Assembly in Cambridge (MA, USA) in 1932. Before that, the IAU Statutes made reference to the International Research Council (IRC, succeeded by ICSU in 1931): Article I, 2 stipulated: "The admission of countries shall be subject to the Regulations of the International Research Council" (Transactions Vol. IV, p. 321). The revised IAU Statutes were then drawn in conformity with those of ICSU (Sect. 4.1). See also Fauque & Fox (2022).

First of all, it is crucial to consider the choice of words: "countries" are said to *adhere* to the IAU, they are not *admitted by* the IAU. This implies a *voluntary act* to join the IAU, and the implication is that any "country" satisfying the legal conditions of Statutes 3(a) or 3(b) should become a member of the IAU, almost automatically. Therefore, the frequently used term "admission" is actually improper in general, strictly speaking it should be "adherence." However, in the case of Taiwan, since the conditions for adherence were not unanimously considered as fulfilled, "admission" here has a different meaning: whether for all countries, until the China case, the adherence was apparently just a formality (more below), for Taiwan it was not, and its "adherence" was put into question and turned into an "admission" issue.

Second, Article 3 says nothing about the adherence procedure. The implication is that the IAU being an association like others, examining adherence (or admission) is left to its Executive Board (here the Executive Committee). Article 3 also says that "A country not adhering to the International Council may adhere to the Union by a National Committee of Astronomy recognized by the Executive Committee of the Union": *what the Executive Committee recognizes is not a "country," but a National Committee of Astronomy.* On the other hand, there is no mention of a "confirmation," or "ratification" procedure by the General Assembly: This is consistent with an "adherence" being approved by the Executive Committee alone, but not with an "admission." Consequently, the Executive Committee (clearly under the guidance of D. Sadler) would want to have its decision at Herstmonceux ratified (for the first time) by the following General Assembly (Berkeley)—at the cost, however, of having to pass via the delicate step of a resolution put to the vote of the Assembly.[56] (More below).

Third, "countries" is put between quotes, because Article 3 also uses quotes when it gives a definition of what is a "country" in the IAU context (last sentence), i.e., not necessarily within political borders. For Taiwan the applicable sentence is *"any dependency in which independent astronomical activity has been developed,"* or, in French *"les territoires ayant une activité astronomique indépendante."* It is interesting to remark that, in the geopolitical context of the late 50s, the terms "Dominions," "diplomatic Protectorates," even "dependencies," refer to a colonial vocabulary, or, at least, imply some form of "dependence" of one country on another.[57,58]

---

[56] Section 3.10.

[57] Even if, in the British Commonwealth system, "dominions" were self-governing (like Canada, since 1867, which had adhered to the IAU as early as 1920), they still had (and have) an allegiance to the Crown.

[58] As a matter of fact, these terms, which were already in use long before the war (precisely since the 1932 General Assembly in Cambridge, USA; see Sect. 3.6), and when India adhered to the IAU in 1946 (i.e., before becoming independent; see Table 2) would disappear in subsequent Statutes. In 1970, the Statutes adopted at the Brighton General Assembly make no reference to "countries," but only to "Adhering bodies" (National Committees of Astronomy and Adhering Organizations (article 6(c)). Also, their Article 7 says: "Adherence of a country to the Union is approved, on the

3. The other issue mentioned in Article 3 is "independent astronomical activity," which in reality, contains two parts: "astronomical activity," and, if so, "independent."

First, according to the Herstmonceux Executive Committee minutes, the question of the existence of any astronomical activity in Taiwan was raised by Kukarkin, supported by Šternbek. Venezuela was not explicitly mentioned, but Sadler is right: As mentioned above, it was the deliberate, inclusive policy of the IAU to accept the adherence of countries with low astronomical activity (once the formal conditions were satisfied).

Table 3.1 mentions the following countries and number of individual members at the time of their adherence, between 1946 and 1958: India (7), Hungary (1), Finland (5), Germany (0—for administrative reasons, corrected later), Venezuela (0), Israel (1), Lebanon (1), Austria (1), Bulgaria (1). In the case of Taiwan, there were actually two individual members (that would be accepted at the Berkeley General Assembly: K.T. Cheng and Tsao Mo); furthermore, I have noted from an inspection of the Transactions, that two more Chinese astronomers from Taipei, not present at Berkeley General Assembly in 1961, but at the Prague General Assembly in 1967, had been IAU members based in Kunming before WWII, at the time of the Stockholm General Assembly in 1938 (Kao Ping-Tse and Shen Zee), so must have been somehow active in Taipei at the time of the Moscow General Assembly.

Of course, given the historical background of Taiwan at that time (in particular the 50-year-long Japanese occupation, see Chap. 2), its astronomers (in particular an obviously small group that had chosen to follow Chiang Kai-shek in Taipei) could only start again from scratch, and their activities at that time appeared to be mainly centered on the history of Chinese astronomy, and solar observations. Fu&Ye note that

> there wasn't such a thing as 'the ROC Taiwan's Astronomical Society'. Its status as a civil entity was only approved by Taiwan's Ministry of Interior one month prior to the IAU's Moscow General Assembly. This was carried out in July 1958 and was at least two months later than its date of application to the IAU. It was legally registered in Taipei, its President was Jiang Bingran,[59] who had left the Mainland for Taiwan in 1946.

But this argument was purely administrative and did not demonstrate that there was no astronomical activity.[60] Actually, the Herstmonceux minutes do mention that Taiwan had sent to the General Secretary a list of its proposed individual members to the IAU "supported by details of their publications" (according to the

---

proposal of the Executive Committee, by the General Assembly; it terminates if the country withdraws from the Union." See the discussion in Sect. 6.3.

[59] Actually, Jian Pigran (Jiang Bingran, 蔣丙然, 1883–1966) was more a meteorologist than an astronomer. He was Professor of the Agricultural School of the University of Taiwan from 1946 to 1966, but he had also been an IAU member. As we can see elsewhere (e.g., Lee 2022), meteorology and astronomy were frequently studied together in the Chinese tradition.

[60] For newly told details about the "astronomical activity" of Taiwan at that time, see Shiang-Yu Wang et al. (2022). We shall return to this important question in the concluding Chap. 8.

Table 3.1 IAU memberships of "divided countries": Germany, China, and Korea (Source: Transactions B)

Year	GA	Total ind. members	Germany	China (Mainland)	China (Taiwan)	Dem. PR Korea	Rep. of Korea
1935	Paris	496	-	6	(Note 1)	(Note 1)	(Note 1)
1952	Rome	809	*Adherence*	2	-	-	-
1955	Dublin	888	0	1	-	-	-
1958	Moscow	1120	65	6	-	-	-
1961	*Berkeley*	*1289*		*Withdrawn* (Note 2)	*Admission* 2	*Adherence* 0	-
1964	Hamburg	1630	*Split between FRG and DRG* (Note 3)		8	0	-
1967	Prague	2009			8	3	-
1970	Brighton	2590			7	3	-
1973	Sydney	3188			7	3	*Adherence*
1976	Grenoble	3805			8	3	6
1979	*Montreal*	*4504*		*Re-joined* (Note 4)	9	3	9
1982	Patras	5200		93	12	3	10
1985	New Delhi	6025		280	14	21	14
1988	Baltimore	6711		307	14	21	19
1991	B. Aires	7301	*Reunification*	292	19	21	32
1994	The Hague	8328		306	23	21	42
1997	Kyoto	8562		369	23	(Note 5)	51
2000	Manchester	8737		368	24		66
2012	Beijing	10904		481	63	(Note 6)	132
2020	(Vienna 2018)	13127		756	81	18	172

Note 1: In 1935, and until 1945, Taiwan and Korea were under Japanese Imperial rule. (See text for details.)

Note 2: "China crisis": China left the IAU for a period of 20 years as a protest against the admission of Taiwan. (Admission approved by EC in 1959, ratified by GA in 1961.)

Note 3: Following the building of the Berlin Wall in Aug. 1961 (construction started on the weekend just before the Berkeley GA).

Note 4: End of "China crisis": China agrees to re-join the IAU, pending final agreement on names of adhering organizations for China (Mainland: Beijing) and China (Taiwan: Taipei). New names of "National members" adopted in 1980 and ratified by GA in 1982: "China, Nanjing" and "China, Taipei".

Note 5: North Korea, admitted at the Berkeley GA at the same time as Taiwan, stopped paying its dues in 1991. Membership terminated on Dec. 31, 1995. Likely due to collapse of USSR in 1991 and resulting massive famine.

Note 6: North Korea reinstated in 2012, also after a period of 20 years.

Statutes, however, the new members could be approved only at the next General Assembly in 1961).

The issue was thus honestly discussed on strictly scientific grounds, but I agree with Sadler that, given the precedents in the inclusive IAU policy since WWII, the level of astronomical activity, whatever it was, could not be used as an argument against the admission of Taiwan. According to the Executive Committee minutes, the other members concurred with this view.

Second, the related question was whether this activity was "independent." Factually, there could be no doubt, since while Taiwan was a geographical area of China (this was never disputed by the IAU; more below), it was physically separated (as an island) and had no administrative ties with it because of the political situation. (Actually, strictly speaking Taiwan had not been under Peking's jurisdiction since 1895.) As long as that situation would prevail, the IAU indeed surmised that the activity should be considered as "independent," with the important justification that, whatever their activity and their numbers, even small, the Taiwanese astronomers formed a group (their Astronomical Society) that was not supported by China, so it had to be considered in the same was as any other recognized group of astronomers (the Astronomical Societies *in*, not *of*, other countries). For the IAU, this was a matter of principle, fully within the terms of Article 3 of the Statutes that had been in force since 1932.

However, in my opinion, this is a misinterpretation of the IAU Statutes, because they were in fact clearly able to handle the two cases together:

(a) *Taiwan:* Statutes 3(a): the "country adheres ... by the organization through which it adheres to the International Council" (i.e., ICSU);
(b) *China:* Statutes 3, sentence before last: "A country not adhering to the International Council may adhere to the Union by a National Committee of Astronomy recognized by the Executive Committee of the Union."

So the problem lied *not* in the conformity (or lack thereof) with the IAU Statutes, but elsewhere: the above "solution," legally acceptable from the point of view of the IAU, was not acceptable by China, because, as we shall see in their reply, they considered at the time that by doing so the IAU de facto agreed (even promoted) that there were "two Chinas," or "One China, one Taiwan," which was unacceptable, just as it had been with the IOC. In other words, the IAU standpoint was that as long as there was no form of mutual recognition between China and Taiwan, there had to be *two* "Adhering bodies" in astronomy, but at that time China and Taiwan (via its official name as "The Republic of China") *both* claimed to represent "One China." (We shall discuss this point over and over again in different contexts below.)

Yet the Chinese had another interpretation: even though there were no administrative ties between China and Taiwan, the activities of astronomers in Taiwan have to be only, also as a matter of principle, under the purview of the Chinese Astronomical Society in Nanking. In other words, *"one country, one Astronomical Society."* This is exactly what Fu&Ye defend:

> It is clearly stated within the Statutes of the IAU that, one country can only have one astronomical representation as a member. Thus, by accepting Taiwan's application was an obvious violation to its Statutes, when China was already represented by the Chinese Astronomical Society in Nanking (Nanjing).

3.7   The Oort-Sadler Proposal

At the Herstmonceux meeting, the various points about accepting the application of Taiwan were thus extensively discussed, with opposing arguments, especially in the light of the "possibility" (the minutes do not use the word "threat") of China might withdraw from the IAU.

Since it soon became clear that a unanimously positive answer could not be reached, according to the minutes, *"After considerable discussion the President put the admission of Taiwan to the ballot; it was agreed to admit Taiwan to membership of the Union by 5 votes for to 2 against."* As demonstrated subsequently, the two negative votes came from Kukarkin and Šternbek; there was thus one abstention, which Fu&Ye claim was by the President.[61] The minutes do not comment on the result, but it was clear that the vote merely reflected the East-West political divide.

At any rate, while the question of Taiwan was considered "settled," the issue remained of what to do in an attempt to convince China not to withdraw. Then Sadler proposed an ingenuous scheme that he had devised with Oort in preparation for the meeting, in the form of a draft resolution submitted to the Executive Committee, based on the interpretation of the Statutes as presented above.

This resolution reads (here typed as a *fac-simile* of the document in the IAU archives, a photocopy of very poor quality):

---

[61] They say this was according to the Statutes, but this is incorrect: Due to the absence of R. Petrie from the meeting, there were only five Vice-Presidents voting, so seven ballots in all. Oort did vote.

```
Whereas there is the possibility of confusion in the descriptions
    variously accorded to the two Republics of China, both of which
    adhere to the Union, and

    in order to indicate unambiguously the two geographical areas in
    which independent astronomical activity has been developed,

    it is resolved that short names, as indicated below, shall be
    used for all purposes of the Union, with the exception of formal
    correspondence and descriptions in which the use of the official
    titles of the two countries is desirable,

1. Official title of the country:    The People's Republic of China
Site of Government:                   Peking, China
Adhering Organization:                National Committee of Astronomy
                                      (Nanking)
Date of Adherence:                    1935
Geographical area in which
Astronomers are represented:          The Mainland of China
Short name:                           China (in French: Chine)

2. Official title of the country:     The Republic of China
Site of Government:                   Taipei, Taiwan
Adhering Organization:                The Astronomical Society of the
                                      Republic of China (Taipei), recog-
                                      nized as a National Committee of
                                      Astronomy by the Academia Sinica,
                                      (Taipei), through which organiza-
                                      tion the Republic of China adheres
                                      to the International Council of
                                      Scientific Unions
Date of Adherence:                    1959
Geographical area in which
Astronomers are represented:          Taiwan (Formosa)
Short name:                           Taiwan (in French: Taiwan)
```

This formulation gives all the technical details and insists on the geographical area (i.e., the "dependence," or the *"territoire,"* in which each membership is valid), and adds "in which the astronomers are represented": resp. "the Mainland of China," and "Taiwan," even adding the geographical name given by the Portuguese, "Formosa." There was no possibility of ambiguity, and it was also obviously impossible to ignore that the "official" name of Taiwan was "Republic of China," i.e., the name under which by Taiwan was known internationally as a member of the United Nations and even of its Security Council!

This resolution was discussed, then also put to the ballot, with the same result: 5 in favor, 2 against.

Then, by an unusual move, B. Kukarkin, supported by B. Šternbek, decided not to accept the result of this majority vote. According to D. Sadler's "personal account":

> The main motion was strongly opposed by Kukarkin, though I am sure that Verlikovski's verbal translation of Kukarkin's Russian was both stronger and lengthier than the original.

In any event, B. Kukarkin (Fig. 3.11) requested to make a formal statement explaining his position.

Fig. 3.11 Boris Vasilyevich Kukarkin (Б. В. Кукаркин), here using a mechanical calculator. He had been Director of the Sternberg Astronomical Institute in Moscow (Credit: Ria Novosti)

The minutes give this statement in full, and in French[62] (here translated):

- Considering that I categorically disapprove of the decision by the Executive Committee of the IAU, voted by 5 against 2, to admit Taiwan to the International Astronomical Union,
- Taking into account that Chinese astronomy, which is developing fast, is represented at the IAU by the Academy of Sciences of the People's Republic of China,
- Emphasizing that there is no independent scientific activity in Taiwan in the field of Astronomy, and thereby the decision of the Executive Committee is in contradiction with the precise meaning of the IAU Statute,
- Being convinced that the adopted decision is detrimental to the interest and prestige of the IAU and to the international collaboration between astronomers,
- I believe it is necessary and even obligatory for me to address the IAU General Assembly of 1961 to ask for a revision of the question of the admission of Taiwan to the IAU,
- In connection with this, I propose that a status quo be observed until the General Assembly of 1961.

To which the President replied

that the matter had been fully discussed and that, in particular, other members of the Executive Committee did not agree with Kukarkin in respect of the independence of scientific activity in Taiwan and the interpretation of the Statutes of the Union: The majority

---

[62]The minutes do not explain why Kukarkin, who needed the help of a translator for expressing himself in English, eventually had his statement written in French (with which Sadler was obviously familiar, as was Oort).

clearly considered that the astronomical activity in Taiwan was independent and that the
Statutes of the I.A.U. did not permit rejection of the application, either on this ground or on
the low level of astronomical activity. He said that Kukarkin was free to raise this matter
again before the General Assembly in 1961, but that he could not accept the proposal to
reverse the decision just taken; this decision to admit Taiwan must stand.

And with that the meeting proceeded normally to the other 29 items of the
agenda.

3.8   The Tumultuous Withdrawal of China

As is well known, China eventually withdrew from the IAU, but the exact circumstances are less well known, and at times rather strange. In short, while China
notified the IAU of its withdrawal in very stark terms, she also wanted to openly
publicize its decision to the world, as revealed by unpublished documents found in
the IAU Archives.

The starting episode is well known and described in Blaauw's book: It is an
exchange of letters between the IAU President, J. Oort, and the President of the
Astronomical Society of the People's Republic of China, Y.C. Chang (whom we
have already mentioned several times). The interested reader will find these letters
reproduced in full in Appendix B1: The extracts selected by Blaauw are seen to be
expunged from many political considerations and statements which, I think, have
their place here since they reflect more faithfully the mood of the times, and the
background in which the IAU–China conflict developed over so many years. (Such
considerations are also referred to explicitly in several documents found in the IAU
archives.)

The exchange begins with a letter by Chang, dated November 20, 1959, to which
Oort replied on December 2. Surprisingly, we (like Blaauw) found no trace of a prior
correspondence from Oort to Chang informing him personally of the Executive
Committee decision to admit Taiwan. In a previous letter to Goldberg, dated October
28, 1959, Sadler mentions: *"So far we have had no reactions from China. A similar
problem was faced by IUPAP[63] and is being faced at present by the IUGG.[64] I hope
very much that China will not withdraw, but the Executive Committee had no choice
but to accept Taiwan's application."* (We will come back later to the issue of the
other scientific unions: Chap. 4.)

In any event, this decision about Taiwan was published in the November 1959
issue (No 21) of the *IAU Information Bulletin*, which some interpret as a *"fait
accompli"* to make the admission of Taiwan forcibly irreversible.[65] (We tend to

---

[63] *International Union of Pure and Applied Physics.* It was founded in 1922.

[64] *International Union of Geodesy and Geophysics.* It was founded in 1919, at the same time as
the IAU.

[65] In his 1976 account, D. Sadler adds that he thought about including the Executive Committee
resolution in this issue, but it was withdrawn at the proof stage by the President, who considered that

disagree with this interpretation: Eventually, the following Executive Committee meeting would take place almost a year later, in July 1960, and there had been a commitment toward Taiwan at the end of the Moscow General Assembly to examine its application "at the next Executive Committee meeting," i.e., in Herstmonceux; since the decision taken there, after the 2–5 vote, was final, why wait to inform the Chinese only via the official channel of the *Information Bulletin*?)

Chang's letter to Oort starts by "*We learn*," without specifying the way he had been informed (perhaps, unofficially, by Kukarkin or some other Soviet informer?), and even questioning the authenticity of the information (which would imply that he wasn't aware of the *Information Bulletin* publication):[66]

> Dear Professor Oort,
>   We learn that the Executive Council Meeting of the International Astronomical Union held in England on September 7, 1959, has adopted the decision of accepting the so-called 'Chinese Astronomical Society' of the Chiang Kai-shek clique for membership in the Union. We are greatly surprised by this decision.

Chang continues:

> As all the people of the world know, Taiwan is an inseparable part of Chinese territory, it is a province of China. It is because of the armed occupation by the American imperialists that Taiwan has not been yet liberated up to the present. The Astronomical Society of the People's Republic of China is the only legal organization to represent China in joining the IAU. The acceptance of the so-called 'Chinese Astronomical Society' of the Chiang Kai-shek clique by the Executive Council Meeting of the IAU is evidently in keeping with the hoax of the American imperialists of creating 'two Chinas'. It is illegal and wrong. Not only is it a trespass on the legal rights of China in an international organization but also an obstruction of the normal development of international scientific cooperation.

He then concludes:

> Therefore, Mr. President, I ask you to clarify the above-mentioned report. Should the report be authentic, I, on behalf of the Astronomical Society of the People's Republic of China, hereby lodge our strong protest with you and insist that the Executive Council of IAU rescind the illegal decision about the acceptance of the so-called 'Chinese Astronomical Society' of the Chiang Kai-shek clique. Otherwise, the Astronomical Society of the People's Republic of China will resolutely and definitely withdraw from the IAU.

So here, we see that the "low activity" of astronomy in Taiwan, which was the central, non-political argument developed by Kukarkin in his Executive Committee statement to oppose its admission plays no role at all. Here, the opposition to the IAU is entirely political and can be summarized by two sentences:

---

the policy had not been approved by the General Assembly. Eventually, this resolution would be presented in identical terms at the Berkeley General Assembly in 1961 (Sect. 3.10).

[66]Blaauw suggests that the *Information Bulletin* of Nov. 1959 was issued "immediately after" the receipt of Chang's letter, implying perhaps a hasty decision, but this seems hardly possible: Chang's letter must have been received at least a week after it was sent (Nov. 20 was a Friday), and Oort's reply is dated Dec. 2 (a Wednesday). Given the usual time constraints of proof-reading, printing, etc., the November publication would have had to be delayed until December (while still being formally the November issue) and likely even after Oort's reply.

- *Taiwan is an inseparable part of Chinese territory, it is a province of China,*[67]
- *The Astronomical Society of the People's Republic of China is the only legal organization to represent China in joining the IAU,*

which is simply the "astronomical corollary" of the first sentence. This is the conundrum the IAU would be seeking to resolve for two decades. The first sentence is still, nowadays, a highly sensitive issue, while the second would eventually find a solution—one which would have to involve Taiwan, which obviously neither Sadler nor Oort had consulted to draw up their resolution in Herstmonceux.

In his reply to Y.C. Chang on December 2, J. Oort, after having presented the resolution adopted by the Executive Committee to consider China's representation in the IAU as *"a temporary, dual one, with one adhering organization located in Peking, and the other in Taipei,"* went on to explain in detail, as diplomatically as possible (and somewhat redundantly) what were the intentions of the Executive Committee, but being very firm on his position: *"The decisions of the Executive Committee cannot be rescinded. In this case the Astronomical Society of the Republic of China (Taipei) has already been informed of these decisions."*

Then Oort wants to clarify the Executive Committee motivations:

I think it is desirable to affirm that the only considerations underlying the decision to admit Taiwan to membership of the Union were scientific, namely the desire to fulfil the objects of the Union by the encouragement of Astronomy in a territory which is, as far as the Executive Committee is aware, not otherwise directly represented in the Union. (...) I must emphasize that the representation covered by the Astronomical Society of the Republic of China in Taipei is limited to those astronomical activities in Taiwan that are at present (and for whatever reason) independent of the guidance, supervision and representation of the Astronomical Society of the People's Republic of China in Peking.

On the issue of the conformity with the Statutes, Oort adds: *"The application from the Astronomical Society of the Republic of China in Taipei formally conformed to the Statutes of the Union and to the Statutes and Directives of the International Council of Scientific Unions, to which the I.A.U. adheres."* Then, he quotes the applicability of Statutes Article 3(a) to the case of Taiwan (which he surprisingly does, without giving a reason, in its French version).[68]

Then comes the issue of the "level of astronomical activity" in Taiwan:

Certainly, the astronomical activity in this territory appears at present to be very limited. But this is no reason for refusing admission. As you may remember, several countries have in the past been admitted, in which astronomical research was only in its very first beginning. Former Executive Committees have felt, as the majority of the present Committee did, that membership of the I.A.U. might contribute to facilitate developing astronomical research. At any rate, the Executive Committee felt they could not discriminate.

To conclude his letter, Oort praises the role of China in the world's astronomy:

---

[67] Since the decision taken by Empress Dowager Cixi in 1885, Sect. 2.2.

[68] The IAU Statutes must be officially published in the Transactions in two versions, English and French, after each General Assembly. Here, Oort refers to the French version of the 1958 Statutes (in Transactions Vol. X, p. 727).

I am certain that I can speak for all members of the Union when I say that the withdrawal of the People's Republic of China from membership of the Union would be a severe loss to Astronomy and a bad blow to international cooperation in the field of science which has so far led the world. I ask you to reconsider carefully your views in the light of this letter and of my assurance that the sole and only purpose for the admission of Taiwan is to help the astronomers working in Taiwan and that astronomers throughout the world are anxious to continue that cooperation with astronomers in the People's Republic of China, for whom we all have high admiration and regard.

Chang replied to Oort in a letter sent on February 5, 1960, a Friday, which arrived at the IAU Secretariat at the Royal Greenwich Observatory about a week later. But in the meantime, surprising events took place which simultaneously made it public.

3.9   An Enigmatic "Radioteletype"

With the mention "PERSONAL and CONFIDENTIAL," in a letter to L. Goldberg dated February 8, D. Sadler writes:

I have been greatly disturbed to have had my attention called by a non-astronomical acquaintance to an American Forces in Europe broadcast (or it could have been a Voice of America broadcast) on Saturday evening 6 February 1960, in which it was stated that the People's Republic of China were considering withdrawing from the IAU on the grounds that it had recently admitted Taiwan to membership. Unfortunately my informant cannot let me have more precise details of the broadcast, as he was only a casual listener. (...) I have no information regarding any publicity that China or the Soviet Union might have given to this matter.

It was a typical Cold War situation: Where was the information coming from? Was this China's official answer, or was there a leak somewhere? Why was the information made public in such an unusual, but at the same time rather confidential way, given the rather narrow spectrum of the listeners? More basically, was it reliable? The broadcast, of American origin, was reported as saying "China was *considering* withdrawing," what was that supposed to mean?

To make matters worse, in the same letter, D. Sadler mentions another problem: *we have heard from the Acting Director General of the Academia Sinica in Taiwan, protesting against the terms of the second resolution* [see above]

passed by the Executive Committee specifying the geographical areas in which independent astronomical activity has been developed. *We do not therefore know at the present moment whether Taiwan is going to continue to adhere or whether China is going to withdraw.* In such a delicate matter as this, which concerns the whole future of the Union, it is intolerable that there should be outside interference, especially by broadcast.

On February 10, there was still no reply from Chang, but Sadler had no delusions. In a handwritten, complementary letter to Goldberg, he seems to be caught by regrets:

I think there is very little chance (1 in 10^N, where N is large) that China will not withdraw. The IAU will have to face a barrage of invectives, possibly a formal resolution condemning the Executive Committee at the General Assembly in Berkeley, but not I think the withdrawal of the USSR; it will undoubtedly survive. But no one can claim that, scientifically, the gain of Taiwan is adequate compensation for the loss of China.

To which Goldberg replied on February 15: *". . . but the Executive Committee rightly concluded that this is not a relevant consideration."*

After some exchange of correspondence, it turned out that the radio broadcast emanated from a reporter having received a communiqué by the official New China News Agency (NCNA) in Peking (best known nowadays as the Xinhua News Agency, 新华通讯社), sent by "radioteletype" (the ancestor of the telex) on "Feb. 6, 1960, at 0800 GMT-W, in English, to Europe and Asia," and simply entitled: "CPR [Chinese People's Republic] withdraws from Astronomical Union."

We found the full transcript of the broadcast in the IAU Archives (Appendix B2). It is based on Chang's reply sent to Oort on the day before (February 5), with extensive quotations from his letter, and comments targeted to the public of "Europe and Asia" to put it in context (and probably also to the USA, sent at a different time?). The text is thus "hybrid": It contains extracts of an official, but in principle private, letter, and explains its whereabouts to the public, as told by Xinhua. As a result, it aims at having more weight internationally than the letter itself, and at preempting any public announcement by the IAU.

As Sadler had predicted, the "barrage of invectives" indeed arrived, perhaps much sooner than he had thought. The preamble of the broadcast is almost identical to Chang's letter of November 20 to Oort asking explanations for the admission of Taiwan and concludes *"In the letter, the Astronomical Society of the People's Republic of China announced its immediate withdrawal from the IAU and all its affiliated commissions, and a discontinuance of all connection with them."*

The communiqué continues:

In his letter to Oort on Feb. 5, Chang Yu-Che said that Oort in his letter 'openly put on a par the great People's Republic of China and the Chiang Kai-shek clique which has long since been disowned by the Chinese people', and with ulterior motives quoted article three of the statutes of the union as 'grounds' for representing Taiwan as a country. Evidently, this is in line with the U.S. imperialist intrigue of creating 'two China's'. (. . .) The Taiwan question is China's internal affair and the Chinese people are determined to liberate Taiwan. We are firmly against the conspiracy of the American imperialists to create 'two China's' and we will oppose it to the end.

Then, the IAU is castigated: *"It is an open interference in China's internal affairs on the part of the IAU to take this as a pretext to deny the right of our astronomical society to represent the astronomical workers in Taiwan. This is utterly intolerable to us."*

In conclusion: *"The admittance of Taiwan to membership is a thing which goes far beyond the scope of 'science', and reduces the IAU to a mere tool in the political intrigue of 'two China's'. This is not only an act of hostility to the Chinese people, but also runs counter to the will of fair-minded scientists in the IAU itself. (. . .)".*

According to Fu&Ye, and probably at about the same time (they do not give the date, nor the reference), *"The official Xinhua news agency of the PRC also published an article titled 'A protestation to the fabrication of the 'two Chinas' plot'. It officially announced the PRC's withdrawal from the IAU."*

Obviously, in their Xinhua press releases, the Chinese authorities linked the admission of Taiwan by the IAU to the conflict with the USA, with astronomy being only in the background. In a letter to Goldberg dated February 18, Sadler explains that Oort and himself agreed to defer the publication of Chang's letter until March 7, in the "forlorn" hope that a letter he had sent to Chang on January 21 might succeed in changing his mind.

But Sadler's February 18 letter contains an important, more optimistic conclusion:

> The withdrawal of China from the Union is a sad blow, but it is by no means a fatal blow to the Union. Our correspondence with Chang has been (allowing for the enforced vituperation against the Americans and anti-communists) very friendly; and Chang has intimated that Chinese astronomers would wish to co-operate in international programmes as in the past.

This point is confirmed by a private letter sent by Chang to Struve later (on June 28, 1960) answering an invitation to attend the Berkeley General Assembly (we will come back to this letter below): In a very friendly and respectful tone (Chang had stayed at Yerkes Observatory when Struve was its Director during his graduate studies in the 1920s, then again after WWII),[69] Chang concludes:

> Of course the withdrawal from membership of IAU does not by any means imply that we stop cultivating the sciences of astronomy in our country. *On the contrary, we hope to be able to make greater contribution to this international cooperative organization when we shall join the Union again sometimes in the future.*

One can imagine that some Chinese political commissar supervised Chang's official correspondence (to Oort, in particular) to make it conform with the government policy and "anti-imperialist" style, thereby not reflecting Chang's personal position, but there can be no doubt about Chang's sincerity and patriotism.

In contrast, Struve, who had invited Chinese astronomers to attend IAU meetings as early as 1955 (Dublin General Assembly), was in a much somber mood, as, in a letter to Goldberg dated February 23, 1960, he was reacting to Chang's February 5 letter and to the broadcast transcript in response to a report that Goldberg had sent to the US National Committee:

> The intemperate language of the broadcast itself and of the quotations from Chang's letter to Oort of February 5 will make it difficult or impossible to heal the breach. (. . .) The IAU's decision to admit Taiwan was, of course, not engineered by 'American imperialists'. The USA has only one representative on the executive committee of the IAU, and even he is not the official mouthpiece of the American government.

Then, he calls for his lifelong experience as an émigré:

---

[69] See Zhou (2022) and Chap. 8.

There have been many upheavals in the world during the past forty years, and millions of people have changed their national affiliations. It would be strange indeed if every country that has lost a part of its population would insist upon representing its émigré's in a scientific union.[70] (. . .) The philosophy that underlies Chang's protest is a dangerous one. If the Union should yield in this case because of the great scientific potential of continental China, it would be plagued for many years with similar 'jurisdictional disputes'.

His conclusion was almost tragic: *"I realize that the Chinese withdrawal may cause other repercussions. We must take that risk. I should prefer to see the Union fall apart than have it submit to threats."*

Eventually, and contrary to all the other important actors of the conflict, Struve would not live long enough to see its happy settlement (he died in 1963).

In the end, the withdrawal of China became official on the IAU side with the publication of the 1960, No.3 issue of the *Information Bulletin*, published in May, three months after the "radioteletype" affair:

Professor Yu-Che Chang, the President of the Astronomical Society of the People's Republic of China, has formally announced the withdrawal of the People's Republic of China from adherence to the Union, in protest against the decision of the Executive Committee to admit Taiwan (the Republic of China) to membership of the Union. As from this announcement astronomers in the People's Republic of China will cease to be members of the Union and therefore will cease to be members of Commissions.

The divorce was pronounced.

3.10   The 1961 General Assembly at Berkeley: A Tortuous Path to Ratification

The next meeting of the Executive Committee (#20) took place in Prague, on July 6–8, 1960, at the invitation of B. Šternbek, the Czech Vice-President who had supported the statement by B.V. Kukarkin protesting against the admission of Taiwan at Herstmonceux. Kukarkin attended, with the same interpreter (P.G. Kulikovsky). Taiwan had been admitted, China had withdrawn as expected, yet there were no hard feelings, no threat to resign, from the two "opponents." This time, the full Executive Committee was present, including Th. Oosterhoff and A. Danjon.

In the minutes of the meeting, the topic of China is addressed in the second item (the General Secretary's report) in only one sentence: *"In connection with the withdrawal of the People's Republic of China from adherence to the Union it was agreed that Chinese observatories should be retained on the Distribution List"* (i.e., of the *Information Bulletin*). For some reason, Blaauw writes in his book that the

---

[70]For reference, recall (Chap. 2) that in 1949 two million Chinese left the Mainland for Taiwan, having a population of eight million at the time. However, Struve somewhat misses the point here, as the Chinese argument rests on authority over territory, not population per se (but of course authority over the population is a consequence).

report mentions that *"letters of protests had been received* [by the General Secretary] *from the Polish and the Bulgarian Academies of Sciences with a request to reconsider the decision,"* but this must have been taken from another document. In any case, Blaauw quotes Sadler addressing the Executive Committee: *"it is pointless for the Executive Committee to reconsider the matter, since it is now impracticable to reverse the action taken as a result of the earlier decision. (...) The General Assembly could, I think, (...) override the decision of the Executive Committee, on a majority vote of adhering countries."* In fact, this statement simply echoed the remark Oort had made after the votes at Herstmonceux, when he said that *"Kukarkin was free to raise this matter again before the General Assembly in 1961."*

That is probably what triggered Kukarkin and Šternbek to think about re-submitting their own Resolutions before the General Assembly in Berkeley. However, the idea was put on the back burner for almost a year.

In the meantime, and in spite of the harsh formulation of the China withdrawal letter, the idea floated around that Mainland Chinese astronomers should be invited as "guests" on an individual basis to attend the Berkeley General Assembly. In fact, even before the Prague Executive Committee meeting (# 20), Struve, in his letter of April 5, 1960, already mentioned, had already taken the initiative to invite his friend Chang. On June 23, Chang replied, with an amicable change of tone and an argumentation previously unheard of:

> Personally, I would like very much to attend the coming Berkeley meeting of IAU and to have the opportunity of meeting you and Prof. Van Biesbroek[71] and other friends. But as a nation we are very sensitive to the Taiwan problem. People in general may think that the liberation or unification of Taiwan and the participation in IAU are two different things. But in these days of thermonuclear weapons and inter-continental missiles, it is not conceivable that national policy should be implemented by force or war.[72] So the alternative is to present the justification of the case to the public opinion. *The withdrawal of IAU is one way of making our point of view well known to the people of the whole world.*

It is too bad that this letter was not publicized and apparently was not brought to the attention of the Executive Committee in Prague. Of course, it would not have reversed the course of events, but it presents the China withdrawal (and the reasons for Chang to decline) under a different, more understandable, no less political but more peaceful albeit realistic, perspective.

In parallel, apparently unaware of the Struve-Chang exchange, and of Chang's negative answer, Oort and Goldberg discussed exactly the same issue (which they had considered together in Prague): Whether or not for Oort (as IAU President) should extend an invitation to Chang to come to Berkeley on an individual basis. On September 6, 1960, in a sobering letter labeled "Private," Goldberg expresses his opposition:

---

[71] George Van Biesbroek, a Belgian-born American astronomer, had been Chang's PhD advisor in the years 1927–1929 at Yerkes (Zhou 2022).

[72] I.e., at the time, the presence of American forces based in Taiwan or cruising in the Taiwan Strait. See also Sect. 4.1.

I now have strong doubts as to the wisdom and even propriety of inviting Chang to Berkeley. I have come to this conclusion after re-reading his letter to you of February 5, 1960 which, as you know, was given very wide publicity by the Chinese themselves. I am quite willing to believe that Chang may have written his letter under duress and that he does not really hold the views he has expressed, but nevertheless there is no dodging the fact that the letter contains a most offensive attack against the government of the United States on the basis of which the Department of State would be quite justified in denying a visa. (. . .) . . . an invitation to someone else might be embarrassing to the I.A.U. in view of Chang's leading role in Chinese astronomy. Under all the circumstances, I would be in favor of dropping the whole matter (. . .).

On September 13, recognizing the strength of Goldberg's argument, Oort replies:

I regret deeply that we shall have no Chinese astronomers at Berkeley, *as I also regretted deeply that the matter of Taiwan's adherence was ever brought up.* I must say that I would not like a situation in which, directly or indirectly, pressure would be exerted by a government on a Scientific Union to abstain from inviting astronomers from a certain country.

More bad news were still to come. As the preparations for the Berkeley General Assembly were well under way, on March 20, 1961, Sadler informs Oort of a new problem: The Kukarkin-Šternbek resolution re-surfaces. *"Unfortunately I this morning received a letter from Sternbek requesting that the question of the membership of Taiwan be placed on the agenda. (. . .) As the matter stands at present, it would be possible for statements to be made* [in front of the General Assembly], *but not in order (unless the President rules otherwise) for any resolution to be considered. If, however, Sternbek formally proposes a resolution on behalf of the Czechoslovak Academy of Sciences* [the adhering organization of Czechoslovakia] *to reach me by 5 April, I propose to include it in the agenda."* Then, a hastily handwritten note is added: *"I have this evening received a similar request from Kukarkin."*

So, after informing Oort, and strictly abiding by the IAU rules, Sadler added to the Agenda two resolutions, written by two Executive Committee Vice-Presidents but submitted formally by their adhering organizations, *against* a decision taken by the Executive Committee after a majority vote, a very unusual, and serious, situation: literally a "vote of non-confidence" against the Executive Committee. The resolution submitted by the Czechoslovak Academy of Sciences was short: *"That the General Assembly revoke the decision of the Executive Committee, in September 1959, to admit Taiwan as an adhering country."* The second resolution, submitted by the Academy of Sciences of the USSR, was longer, essentially repeating the Kurkarkin-Šternbek statement at Herstmonceux, but updated with the withdrawal of China, and concluding: *"The General Assembly resolves to cancel the decision to admit Taiwan, as adopted by the Executive Committee on 8 September 1959."*

As per the IAU rules, the two resolutions were published in advance in the *Agenda and Draft Reports* of the General Assembly (under item 10: "Resolutions submitted by Adhering Organizations or National Committees of Astronomy"; there were two more resolutions, of a scientific nature, by Australia and Belgium), and distributed to all participants. The vote was due to take place on the last day of the General Assembly, in the "Admission of new National Members" session: For such a vote, the constituency is not the totality of the participants, but the officially appointed "National representatives" of the IAU adhering countries (35 in Berkeley):[73] "one country, one vote."

Upon receiving the *Agenda and Draft Reports,* O. Struve, in a handwritten letter to L. Goldberg dated July 2, 1961, expressed his concerns: *"the resolutions [by the USSR and Czechoslovakia] were implying a 'vote of no-confidence' in the present Exec. Comm., and this could easily be disastrous for the future existence of the Union".* But in this reply (July 6), Goldberg was quite reassuring: *"In Sadler's opinion, the Soviet-Czech resolutions will be supported by about six countries and there may also be an equal number of abstentions, but the resolution will be easily defeated."*

On the other hand, since the General Assembly was scheduled to begin on Tuesday, August 15, and end on Thursday, August 24, the National representatives present in Berkeley had 9 days to discuss the issue between themselves and with other participants. To simplify the vote, during the Assembly, the official representatives of Czechoslovakia and the USSR were asked by the General Secretary to combine the two resolutions into a one, the Soviet resolution, called "the motion" in the Proceedings of the General Assembly.[74]

The voting procedure, discussed within the Executive Committee and which D. Sadler mentions *"was agreed, due to the generous cooperation of Kukarkin and Sternbek,"* took place as follows:

- Presentation of the motion and its history, by the General Secretary;
- Statement read by the President, as prepared by the Executive Committee: *"On behalf of the Executive Committee, I formally move the rejection of the motion and ask representatives of adhering countries to vote against the motion"*;[75]
- Justification and explanations for this recommendation (based on the minutes of the Herstmonceux meeting);
- Open discussion;
- Ballot.

Out of the 1289 IAU individual members recorded in 1961, 711 (40%) were eventually present at the Berkeley General Assembly, coming from 38 countries (adding the ones that would be admitted: Taiwan, Brazil, Turkey). The largest attendance of course came from the USA (273 participants), with 42 from the USSR (Fig. 3.12). The Business Session discussing the resolutions took place as planned, on August 24, 1961. The Transactions (vol. XI B), edited by Sadler, give a very detailed account of the session (three dense pages), worth reading attentively. It

---

[73] Not counting Brazil, a member since 1922; it re-joined the IAU at Berkeley (probably after a period in which it couldn't pay its dues). The list of countries is given in the Transactions Vol. XI B (1962), pp. 19–26; missing ones (representative not present) were: Bulgaria, Lebanon, and Romania.

[74] Transactions Vol. XI B (1962), pp. 33–35.

[75] Note that the statement was "prepared," not "approved" by the Executive Committee -otherwise Kukarkin and Šternbek would have proposed to vote against the resolutions they had submitted themselves!

Fig. 3.12 Adlai Stevenson, J.F. Kennedy's Ambassador to the United Nations, giving his opening address in front of a huge assistance, at the XIth General Assembly in Dwinelle Plaza, in the sunny Berkeley Campus, on August 15, 1961. (From the General Assembly daily newspaper "*IAU News Bulletin*"; © IAU)

is most unfortunate that this volume of the Transactions cannot be downloaded at present (due to an unsolved copyright issue),[76] so we hereby select the most relevant extracts:

> The General Secretary (. . .) pointed out that it was quite possible to adopt different points of view on this matter and that some of these are incorporated in the resolution as proposed by the U.S.S.R. Academy of Sciences. However, the majority of the Executive Committee had taken a rather different view, which would be presented to the General Assembly in a statement by the President (. . .)

After reading the above statement, asking the representatives to reject the motion, the President went on to read the text prepared by the Executive Committee:

> "There can be no question that Taiwan qualifies for adherence to the Union under the terms of the Statutes of the Union." (. . .) "In recent years, the Union has been glad to welcome the adherence of other countries in which astronomical activity is on no higher level. There is no

---

[76] See https://www.iau.org/publications/iau/transactions_b/. Hopefully, this problem will be fixed in the future.

evidence that this astronomical activity is not independent; it is directed by a properly constituted National Committee of Astronomy."

Then alluding to China's withdrawal:

"it must be made absolutely clear that the decision to withdraw was made by China alone, for reasons that have nothing whatsoever to do either the Union or with Astronomy. The Executive Committee was, however, informed the China might withdraw if Taiwan were admitted, and for that reason approved a resolution which it was hoped would enable China to maintain its adherence to the Union."

This resolution reads: (follows the scheme presented by D. Sadler at the Herstmonceux meeting, see Sect. 3.7).

And concluding:

The Executive Committee can only express the hope that circumstances will soon change in such a way as to enable China to resume its adherence to the Union, and so to play its full part, formally, in international co-operation in Astronomy. We have noted the great strides that are being made in China, and have learned with pleasure from Chinese astronomers that international co-operation will continue informally even though China no longer adheres to the Union.

Then, the discussion was opened by the President. Kukarkin raised his hand and commented that *"at this stage he wished to dissociate himself from the statement that had been made* [by the President] *on behalf of the Executive Committee."* Rather surprisingly, apparently no further comments came from the audience, and the motion was put to the ballot.

The motion was rejected (following the recommendation of the Executive Committee) by 24 votes against, 5 in favor, with 3 abstentions and 1 absence because of illness: *"The President declared that the motion was lost."*

Doing the math, and as predicted by Sadler, the result could be expected: There were five countries present belonging to the Eastern bloc (Czechoslovakia, Hungary, Poland, the USSR, and Yugoslavia; Bulgaria and Romania were absent); the three abstentions might have included Egypt (this was after the Suez canal crisis) and Indonesia (a leader of the "non-aligned countries")? In any case, the votes were anonymous, but clear: The resolution submitted by the USSR was a lost cause. In the same way as the vote at Herstmonceux, the result clearly followed the East-West division of the world ...

The session on Taiwan, however, was not completely over. After giving the result of the vote, the President bade farewell to Chinese astronomers, by *"regretting their absence from the General Assembly and sending every good wish for astronomical research in China."* The proposal to send them a message *"was received with acclamation."*

Then A. Blaauw, the representative of the Netherlands (who would play a key role as IAU President on the settlement of the China issue 15 years later),[77] asked about the usage of official and unofficial names in the IAU Proceedings, as they had appeared on the President's statement before the vote (i.e., Sadler's Executive

---

[77] Chapter 7.

Committee resolution). As the matter was not on the Agenda, the President asked the General Secretary for a statement. Actually, this question concerned Taiwan, not China: The Academia Sinica in Taipei had sent a letter to the General Secretary *"formally requesting that the question of the name used for their country in the Proceedings of the Union be submitted to the General Assembly, as they do not approve of the resolution adopted by the Executive Committee."* In the absence of any request for discussion by the National representatives, the matter was not discussed further.[78]

Not mentioned in the Transactions, but explained by D. Sadler in his "Personal account" of 1976, is the real reason why the Academia Sinica did not approve of the wording of the Resolution: In fact, they wanted the "short name" for the "official title of the country," i.e., "The Republic of China," to be also "China," not "Taiwan," which was clearly impossible. In practice, the Executive Committee, at its meeting (# 22) on August 24 concluding the General Assembly, decided that *"care would be taken to ensure that the agreed [by the Executive Committee] short names were always used in the future"* [resp. "China" and "Taiwan"], adding *"It is necessary, however, to use the description 'The Republic of China' in certain limited contexts."* This point was a sticky one, which re-surfaced during the last steps of the "reconciliation," nearly 20 years later, as discussed below (Chap. 7).

In retrospect, the vote at Berkeley on the admission of Taiwan, which was triggered by the submission of resolutions that, potentially, could have overthrown the Executive Committee (at least its Officers, namely the President and the General Secretary, who would likely have had to resign) had the opposite result: With the admission of Taiwan de facto ratified by the General Assembly (whereas, as noted before, this was not a statutory obligation) not unanimously but with a very large majority, the legitimacy of the Executive Committee, as the main IAU institution, came out reinforced—until it was challenged 15 years later (see Sect. 5.7).

3.11   New Cold War Challenges for the IAU at Berkeley

In the background, new Cold War clouds were quickly piling up at Berkeley that would test the diplomatic skills of astronomers and the IAU Executive. Dominated by the China–Taiwan–IAU crisis, the Berkeley General Assembly turned out in fact to be a tipping point for the relations between the IAU and geopolitics.

Not only had China withdrawn a few months before in highly contentious conditions, but on the weekend just before the beginning of the General Assembly, in the night from Saturday, August 12 to Sunday 13, East German leaders launched the construction of the Berlin Wall, pitting the USA and the USSR directly against each other. Maybe this event influenced the General Assembly participants so as to follow the Executive Committee in their vote about Taiwan. Nevertheless, as it

---

[78] I skip here some procedural details that appear in full in the Transactions Vol.XI B (p. 35).

happened, this key geopolitical development coincided with the election of a Soviet astronomer, V. Ambartsumian, as J. Oort's successor (he had been proposed as President by the IAU "Special Nominating Committee" several months before the Berkeley General Assembly), with D. Sadler as General Secretary, L. Goldberg and B. Šternbek as Vice-Presidents, all for a second term.

Because of the Wall (and increased controls and barbed wire all along the border between the two countries), the circulation between citizens of "the two Germanies" that were created after the Berlin blockade of 1948–1949 became extremely difficult. The German astronomers had succeeded so far in remaining members of a single Society of Astronomy across the East–West border (the century-old *"Astronomische Gesellschaft"*), thus being (since 1951) a single National member of the IAU, representing Germany as a whole. But now they were forced to split between two adhering organizations: the Democratic Republic of Germany (East Germany) kept being represented by the *Astronomische Gesellschaft*, and a new organization was created to represent the Federal Republic of Germany (West Germany): the *"Rat Westdeutscher Sternwarten"* (RDS), or the Council of West German Observatories. The IAU now had *two* German adhering bodies! The situation could only go back to normal 30 years later, in October 1990, following the fall of the Berlin Wall and the subsequent German reunification.[79]

It is also at the Berkeley General Assembly that the Democratic People's Republic of Korea (North Korea) adhered to the IAU—*precisely during the same session* in which the admission of Taiwan was ratified by the General Assembly. This event, involving another "divided country," went however unnoticed (in fact no North Korean astronomer was present at Berkeley, in all likelihood because of the impossibility to obtain a visa), and it is not even mentioned in Blaauw's book. But it should be considered very significant in the heated context of the admission of Taiwan: While there was unexpectedly more than "minimal astronomical activity" in North Korea, it had just been admitted by ICSU, and its adherence to the IAU was in conformity with the Statutes article 3(a), just like Taiwan. In spite of the USA–China face-off during the Korean War, there was no pressure from the US State Department on the IAU to admit the "Republic of Korea" (South Korea), even though it had been accepted as an ICSU member at the same time as North Korea. The adherence of South Korea to the IAU would take place freely much later, in 1973. But contrary to Germany, Korean reunification, once envisaged, has still not happened today.[80]

---

[79] R. Wielen (2019) and (2022).
[80] Lee (2022).

Chapter 4
China vs. Taiwan: A General Problem for Scientific Unions

4.1 The International Scientific Unions and China in 1958

At the end of the Berkeley General Assembly in 1961, the IAU had "gained" a new member, Taiwan, but at the cost of China withdrawing voluntarily because it considered Taiwan as "province of China." It was thus a divorce-like situation, which developed in a painful succession of events, overshadowed by suspicion of political interference by the USA.

It is now time to put this "divorce" in perspective, by comparing the situation of the IAU and that of other international scientific unions with respect to their relations with China. To do so, it is mandatory to examine first the Statutes and, if applicable, resolutions concerning membership, starting with the "Unions of unions," ICSU (standing for the *International Council of Scientific Unions*). ICSU was born in 1931, succeeding the *International Research Council* (IRC) formally established in 1919 in Brussels,[1] and merging in 2018 with the *International Social Science Council* to form the current ISC (*International Science Council*). Its Statutes evolved with time, but here we are concerned with the Statutes and resolutions in force in the time frame of the Berkeley meeting.

The ICSU Statutes of 1931 (which remained valid until 1994) stipulate in particular that (Greenaway, p. 190):

> (III) The International Council of Scientific Unions consists of a national scientific organization from each country which has adhered to the Council and of the International Unions.
> (IV) A country may join the International Council either through its principal Academy, or through some other national institution of association of institutions, or, in the absence of these, through its Government.

---

[1] Fauque and Fox (2022).

T. Montmerle, Y. Zhou, *China and the International Astronomical Union*, Historical & Cultural Astronomy, https://doi.org/10.1007/978-3-031-01787-2_4

Equally important was the ICSU Policy Resolution on "Political Non-discrimination of Scientists" (the first in a series of many along the line of the "Free Conduct of Science"), adopted by its 8th General Assembly, Washington DC, in October 1958,[2] in particular items 1 and 2 (Fig. 4.1).

The adherence of the IAU to ICSU meant conforming to both the ICSU Statutes and its Policy on Political Non-discrimination of Scientists. The IAU had accordingly approved a revision of its Statutes at its IVth General Assembly in 1932 (Cambridge, USA). As we have shown in the preceding Section, this was a central argument in the process of admitting Taiwan.

The ICSU Statutes also illustrate a fundamental difference with other international organizations: It is a "non-governmental," meaning a non-political, organization, because "it has no implication with respect to recognition of the government of the country concerned." In contrast, all UN-affiliated organizations, and in particular, for what concerns scientific unions, UNESCO (the "S") are "governmental," with a one-to-one correspondence between a seat in the UN and membership in these organizations: We will also see (Chap. 6) how this was going to deeply affect the China–Taiwan situation.

The General Assembly,

In keeping with the purely scientific character of ICSU approves the following statement:

1 To ensure the uniform observance of its basic policy of political non-discrimination, the ICSU affirms the right of the scientists of any country or territory to adhere to or to associate with international scientific activity without regard to race, religion or political philosophy.

2 Such adherence or association has no implications with respect to recognition of the government of the country or territory concerned.

3 Subject only to the payment of subscriptions and submission of required reports, the ICSU is prepared to recognize the Academy, Research Council, National Committee, or other bona fide scientific group representing scientific activity of any country or territory acting under a government, *de facto* or *de jure*, that controls it.

4 Meetings or assemblies of ICSU or of its dependent organisms such as its Special Committees and its Joint Commissions should be held in countries which permit participation of the representatives of every national member of ICSU or of the dependent organisms of ICSU concerned, and allow free and prompt dissemination of information related to such meetings.

5 ICSU and its dependent organisms will take all necessary steps to effect these principles.

Fig. 4.1 The ICSU Policy Resolution on "Political Non-discrimination of Scientists" adopted in 1958 (Reproduced from Greenaway, p. 94)

---

[2] This is the ICSU General Assembly (following the Moscow IAU General Assembly), which had been attended by Oort and Sadler, and in which Taiwan was admitted (see Chap. 3)

China was admitted to ICSU in 1937, when its capital was Nanking, through the Academia Sinica (founded in 1928). According to Greenaway (p. 89), ICSU was still based in Nanking at the time of its 5th General Assembly (Copenhagen) in 1949, but it had moved to Peking (Mao's new capital) at the time of the next one (Amsterdam, 1952), instead of Taipei. In other words, we have no details but at some point the ICSU Executive decided to leave the original Academia Sinica for a new adhering organization (called the Scientific and Technical Association of the People's Republic of China in 1958; today the Chinese Association for Science and Technology, CAST).[3] In other words, after 1949 the Republic of China (Taiwan) was not a member of ICSU, but it was admitted in 1958 at its 8th General Assembly in Washington, the same that would issue its Policy Resolution on Political Non-discrimination of Scientists—and the same year the IAU General Assembly took place in Moscow.[4] As a result, the People's Republic withdrew from ICSU (under circumstances that would need to be clarified; this withdrawal is not mentioned in the Executive Committee minutes of the Herstmonceux meeting a year later, so it may have taken some time), as it did from the IOC.[5]

Now let's take the example of two of the Unions[6] founded in 1922: IUPAP (*International Union of Pure and Applied Physics*) and IUPAC (*International Union of Pure and Applied Chemistry*).

China became a member of IUPAP in 1934. Unfortunately, it seems that no "History of IUPAP" exists in any way similar to Blaauw's *"History of the IAU."* Their website gives however a summary of the main events for each General Assembly over the period 1922–1992 (70th anniversary), alternatively in English or in French.[7] China's admission occurred in 1934 (4th General Assembly in London). Then, there is a 10-year gap between 1937 and 1947 (obviously due to WWII). In 1957, the USSR became a member, and *"the 9th General Assembly*

---

[3]Unfortunately, as we will see frequently in the exchange of letters and documents, and somewhat surprisingly, the IAU and other non-Chinese correspondents use almost interchangeably "Academia Sinica, Peking," or "Taipei," or nothing, or else "Academy of Sciences," etc. in a similar way, which is at times confusing. It is also confusing, on the Chinese side that they continue to use "Academia Sinica" for some institutes, e.g., "Peking Astronomical Observatory, Academia Sinica." See also Sect. 7.3. Officially, the original Academia Sinica of 1928 moved to Taiwan in 1949 (the name having been conserved), and some elements that remained on the mainland gave rise to today's "Chinese Academy of Sciences" (CAS), with 1949 as its actual creation date.

[4]Actually, the ICSU decision to admit Taiwan must have been taken in advance of its General Assembly in Washington, which took place in October, whereas the IAU General Assembly in Moscow had taken place before, in August. This may explain the haste with which Taiwan had sent its application to the IAU (see Chap. 3, and also below). This "inverse chronology" also happened concerning the adherence of North Korea to the IAU at the Berkeley General Assembly in August 1961, since it was formally admitted to ICSU only on the following September, at its 9th General Assembly in London.

[5]Section 3.3.

[6]For short, we shall write "Unions" (with a capital "U") instead of "International Scientific Unions"; as a rule, "Unions" also implies "belonging to ICSU."

[7]http://iupap.org/wp-content/uploads/2013/04/history.pdf

[Rome] *learned that Pekin* [sic] *would be seeking admission to IUPAP."* The report of the 10th General Assembly (Ottawa, 1960) gives puzzling conclusions (in French, here translated *verbatim*):

- East Germany is admitted, the adherence of countries formerly members (Pakistan, Rumania, China (Formosa)) is ratified.
- Mainland China is admitted and subsequently refuses its admission because of the participation of the Republic of China.

This suggests a scenario in which China left IUPAP at some point (perhaps for being unable to pay its dues), but came back in 1960 (although the summary mentions "admission," not "re-admission"). In any case, the net result is that Taiwan (here called Formosa) was admitted in 1960, causing China (former member or new member?) to withdraw on the spot.

The context was different for IUPAC. From IUPAC records, it turns out that China was not a member before 1949, but the Chinese Chemical Society, based in Taipei (Taiwan), was admitted as a National Adhering Organization in 1959 (the report of the then President of IUPAC, S. Stoll, says *"after having paid its dues, so there is no obstacle in admitting Nationalist China as a member country")*.[8] In this case, the People's Republic did not have to withdraw, simply because it was not a member of IUPAC, and apparently had not asked for admission. But what is noteworthy is that Taiwan was admitted to IUPAC at about the same time as it was to IUPAP.

Another important point to consider is the close relationship between ICSU and UNESCO, established as soon as UNESCO was founded, in 1945 (see Chap.5 of Greenaway). In essence, both organizations were formally independent (ICSU being non-governmental and UNESCO a UN organization—the "O"—hence governmental), but ICSU acted frequently as an advisor of science topics—the "S"—and even as an intermediary to fund its Unions, following a convention signed in November 1946.[9] In fact, UNESCO initially provided for a very large fraction of the ICSU budget: from 100% at its foundation in 1946, to still ~60–70% in the 1950–1960 period (Fig. 4.2). In particular, it should be remembered that UNESCO played a very important political and financial role in implementing the International Geophysical Year (1955–1958), which provided the main motivation for the USSR to join ICSU in 1955.

In parallel, after the war many new Unions were created, in addition to the "historical" ones such as the IAU, IUPAP, IUPAC, etc. According to Greenway (p. 90), at the ICSU 6th General Assembly of 1952 in Amsterdam, there were 11 new ones, often created with the patronage of UNESCO. In addition, by construction so to speak, the "Republic of China" (Taiwan) was at that time a member of UNESCO, while the "People's Republic of China" (Mainland China) was not. The opposite was true for ISCU, as we just saw, so this uneasy situation might have been a problem.

---

[8] D. Fauque, private communication.
[9] See UNESCO (2006).

Fig. 4.2 UNESCO subvention as a percentage of the ICSU budget [After Fig. 3 of Greenaway (1996)]. The percentage was over 70% when Taiwan was admitted to ICSU and other scientific unions. It had declined to about 40% when UNESCO asked ICSU and its Unions to expel Taiwan, after its expulsion from the UN in 1971 following the admission of China (Chap. 6)

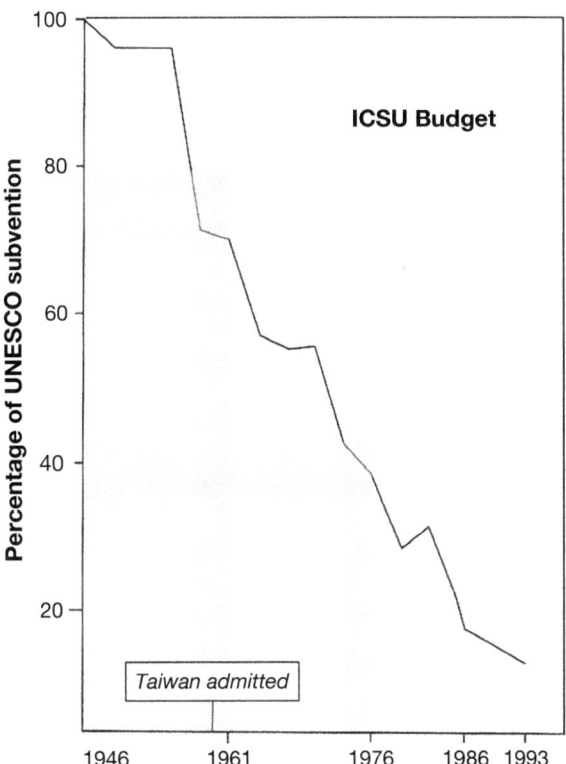

These Chinese memberships would swap in 1971, when China (the "People's Republic of China") became a member of the UN, and Taiwan (the "Republic of China", one of its founding members) was expelled: For ICSU and the IAU, this would have serious consequences (see Chap. 6; also Fig. 4.2).

Thus, an important consideration is to know which Adhering Organization (if any) first represented Chinese scientists in a given Union, and when. However, apart from a few cases to be discussed later, similar information seems hard to find for other Unions (which, in the best cases, only list the admission dates corresponding to the post-1979 era, see below), and it is beyond the scope of this book to make further detailed investigations.

What emerges from the examples above, however, is that Taiwan appears to have applied to several Unions (some with strong links with UNESCO, hence with the UN, including ICSU itself), at about the time of the ICSU 1958 General Assembly in Washington.

Exactly relevant to this period is therefore the disturbing testimony of D. Sadler in his "Personal account" of 1976. After the Moscow General Assembly in August 1958, he and J. Oort went to Washington in October as IAU representatives. They wanted in particular to clarify the status of Taiwan, in preparation for the Executive Committee meeting that would be held in Herstmonceux the following year. D. Sadler writes:

At the ICSU meetings, October 1958. Although not strictly relevant, the question of the two Chinas was, of course, brought up at the meeting of the Executive Board of ICSU at the Academy of Sciences, in Washington D.C. (almost in the shadow of the State Department) in October 1958. As far as I can remember, it was not discussed during the General Assembly. Jan Oort and I represented the IAU. We were immediately appalled by the list of adhering countries, boldly announcing the membership of China (The Republic of China) with adhering organization the Academia Sinica, Taipei. We discovered that several other Unions (to which China adhered) had recently received applications from Taiwan, and I was informed that the *(illegible)* letters of application had been drafted in the U.S. State Department by a scientist (who shall be nameless), who was actually a member of the ICSU Executive Board. I have never had any reason to doubt the accuracy of that information. During the meeting of the Board, Laclavère[10] (IUGG), Oort and I repeatedly attempted to raise the question of the representation of China, but were always faulted by the President of ICSU (L.V. Berkner);[11] also there was a majority (mainly from the ranks of the non-Union representatives) who clearly did not want the issue discussed in Washington.

It thus seems that ICSU itself, perhaps because of its strong ties with UNESCO, was pushing for the admission of Taiwan in its member Unions, after having initially chosen to follow the People's Republic of China, and leaving them with the responsibility of dealing with its withdrawal thereafter. Given the credentials of its American President Berkner, in particular the organization of the UNESCO-sponsored IGY (see above), one can speculate that the initial idea came from him, which would explain why the US State Department played a role at a high level—different, however, from what it was in the case of the IAU.[12] It is puzzling that Taiwan's campaign of applications came precisely at the same time ICSU was promoting its Policy on Political Non-discrimination of Scientists. However, this policy was rather a long-term goal, whereas the admission of Taiwan seems to have served short-term geopolitical interests.

Why, then, was this particular moment chosen? According to Fu&Ye (and not mentioned by Blaauw, not even in D. Sadler's account), this was linked with the conflictual situation over the Taiwan Strait.[13] They write:

---

[10] Georges Laclavère, geodetist. Long time Treasurer of IUGG (International Union of Geodesy and Geophysics, founded at the same time as the IAU (http://www.iugg.org/IUGG90Years/EOS_preprint.pdf) (Fauque & Fox, 2022). He became the ICSU Treasurer from 1961 to 1968.

[11] Lloyd Viel Berkner, outgoing ICSU President (elected in 1955), was an American physicist and engineer. He was a member of the President's Science Advisory Committee (PSAC), which was created on November 21, 1957, by D. Eisenhower (then US President) as a direct response to the launch of the first Soviet artificial satellites. His main achievement was to set up and supervise the organization within ICSU of the International Geophysical Year (IGY: 1955–1958). See Hales (1992).

[12] The IAU (L. Goldberg) was in contact with W. Brode, who was the science advisor to the Secretary of State J.F. Dulles, because of the question of visas to attend the Berkeley General Assembly. See Chap. 3.

[13] This connection is also explicitly mentioned about China and the IOC by Espy (1979), p. 63.

In December 1954, Washington signed the 'Mutual Defense Treaty between the United States of America and the Republic of China' with Chiang Kai-shek. This development further aggravated the long-term standoff and tensions across the Taiwan Strait. Since then, the Taiwan issue had become a major dispute between the US and China.

As a matter of fact, after a first crisis during which the above treaty was signed, a second crisis erupted on August 23, 1958 (i.e., right after the end of the Moscow General Assembly), with fights between the People's Liberation Army artillery and the Republic of China Army (ROCA), supported by the US Seventh Fleet.[14] For several weeks, with the Korean War just 5 years back, China and the USA were again almost at war.

In this dangerous context, pending further investigations, the simultaneous applications of Taiwan to many scientific unions might then be understood as a "peaceful" effort to protect Taiwan's government, but perhaps also, from the ICSU point of view, to invest in the future of the island and its population.[15] In any case, the resulting decision by China to withdraw from these unions, which had no problem in admitting her (contrary to the USA which opposed its admission to the UN), was her own political choice, the price to pay for considering Taiwan as "part of China."

4.2   The Aftermath of the "China–Taiwan Crisis": The IAU vs. Other Unions

In Chap. 3, we have abundantly discussed the events and debates that took place during (and offstage) the IAU General Assemblies, even quoting numbers: On the order of 800 participants for the Berkeley General Assembly for instance (and about the same for the Moscow General Assembly). From Sect. 4.1, the casual reader may get the impression that, for other unions (ICSU, IUPAP, IUPAC, etc.), comparable General Assemblies are being held at regular intervals. But the reality is totally different: The IAU (even today) *is the only major international scientific union that has individual members.* As we have seen, National members also exist (see the Statutes), but other unions all have only "representative members": ICSU, for instance, is composed not only of all its unions, but also of national bodies (countries) and scientific associates (usually smaller unions, specialized committees or councils). When their General Assemblies meet, the participants are in reality delegates with specific mandates (like voting instructions) from their respective institutions, i.e., they represent their colleagues as members of these institutions, having to follow officially recognized processes. In short, their General Assemblies

---

[14]The source of conflict was the status of the Quemoy (Chin-men, or Kinmen, 金門) and Matsu (Lienchiang county, 連江縣) groups of small islands right off the coast of Mainland China but belonging to Taiwan (see Fig. 2.1). A *status quo* holds since this crisis.

[15]Much later, but shortly after the academic bodies of Taiwan and China both became members of ICSU in 1982 (see below), the Taiwanese chemist Yuan Tseh Lee (李遠哲, Li Yuanzhe) was awarded the Nobel Prize, in 1986 (About his name, see also Sect. 2.5).

are essentially debating *science policy*, in each branch of science, not the science itself.

In contrast, IAU General Assemblies are (apart from administrative "Business Sessions") full scientific meetings, with astronomers grouped in Commissions, Working Groups (and now large thematic Divisions) discussing the latest advances in their fields.[16] In other words, the IAU is in reality *the community of the world's astronomers.* That is why it is so inclusive that it accepts small countries and/or countries with (provisionally, it is hoped) so-called "low astronomical activity," as Taiwan was said to be at the time of its admission. This is fundamental to understand the "personalized" response of the IAU to the admission of Taiwan, and the subsequent withdrawal of China, over a 20-year time span, and the lifelong implication of some *"astronomers as diplomats"*[17] willing to mend fences.

In contrast, and pending a more detailed study, in the aftermath of the ICSU General Assembly in Washington, there was not much uproar within ICSU and the other unions following the withdrawal of China (see, e.g., Greenaway, p. 101). The situation would change later, even generating an internal crisis within ICSU itself.

As for the IAU, as predicted by D. Sadler in the letter to L. Goldberg following the withdrawal of China, quoted above: *"The IAU will have to face a barrage of invectives, (...) but it will undoubtedly survive."* While the "Taiwan-China" crisis left many scars, other topics and discoveries also captured the attention of astronomers during the 20 years of absence of China from the IAU, if only because of the advent of the space age, that had begun just before the Moscow General Assembly with the launch of *Sputnik-1*, the Earth's first artificial satellite.

---

[16] Montmerle (2019), in IAUS 349, pp. 289–324.

[17] This is the title of Montmerle and Fauque (2022).

Chapter 5
Living in Separate Worlds

5.1 The "Ice Age"

At the end of the Berkeley General Assembly, on August 24, 1961, a new Executive Committee took office. As mentioned above, the President was a Soviet astronomer (Viktor Ambartsumian, former Vice-President, who would preside over ICSU in 1968 and 1970), with D. Salder starting a second term as IAU General Secretary, and J. Oort as Advisor.[1] China was not anymore on the agenda of the Executive Committee meeting (#22) concluding this memorable Assembly.

The next Executive Committee meeting (#23) took place the following year, from August 30 to September 3, 1962, in Erevan, the capital of Armenia (then a Soviet Socialist Republic), located in the valley leading to the Byurakan Observatory, which Ambartsumian had created in 1946. A notable request was from the Democratic Republic of Germany to adhere to the IAU (see Sect. 3.10). But there were no news from China. Then, Liège (Belgium) was the place of the next Executive Committee meeting (#24): This time, the Federal Republic of Germany informed the Executive Committee that steps were being taken for its adherence. Apart from technicalities, the case would be settled in 1964 without problem at the next General Assembly in Hamburg (what a coincidence!), since this was a coordinated effort by the two "divided countries," neither claiming to represent Germany as a whole.[2] But again no mention of China (or Taiwan).

In 1965, at the Executive Committee meeting (#27) hosted at Nice Observatory by its Director, the new General Secretary Jean-Claude Pecker (with Pol Swings as new President, and D. Sadler still a member as Advisor), the issue of China was raised. The minutes report:

---

[1] The Vice-Presidents were as follows: *First term,* Yusuke Hagihara (the first Japanese Vice-President since Japan had re-joined the IAU in 1952; 1897–1979) and Guillermo Haro (Mexico; 1913–1988); *Second term:* L. Goldberg, R.H. Stoy, R.M. Petrie and B. Šternberk.

[2] See Wielen (2022).

© The Author(s), under exclusive license to Springer Nature Switzerland AG 2022   69
T. Montmerle, Y. Zhou, *China and the International Astronomical Union,* Historical & Cultural Astronomy, https://doi.org/10.1007/978-3-031-01787-2_5

Members of the Executive Committee had expressed concern over the undesirable situation regarding communication with astronomers in the People's Republic of China and proposed that means be sought to improve this situation. They felt that although no hope existed for any sort of official affiliation of the People's Republic of China to the IAU at the present time, certainly some unofficial approaches might be made. The various suggestions put forward included: i) circulating (but not in the name of the IAU) a list of names of Chinese astronomers to Western astronomers who might initiate individual contacts; ii) inviting (through the Czechoslovak Academy of Sciences) Chinese astronomers to visit Prague at a time to coincide with the IAU General Assembly (to be held in 1967);[3] iii) inviting [through IAU Commission 5 (Documentation & Astronomical Data)] representatives from China to present a review of astronomy in their country at the IAU General Assembly; iv) informing the President of Commission 38 on the Exchange of Astronomers of this problem.

A brief follow-up of these suggestions appears in the minutes of the following Executive Committee meeting (#28) in Prague (1966): *"The General Secretary reviewed briefly the content of his consultations with Professor Mao Lin Tcheng [4] of Academia Sinica* [Peking] *as to the relationship of the People's Republic of China to the Union, and pointed to the negative results they had led to."*

However, outside of the IAU, personal initiatives by astronomers did take place. First and foremost, there was a continuing collaboration with Soviet astronomers during the 1950s, at the height of the Sino-Soviet relations. There were many academic exchanges,[5] and six Chinese astronomers had attended the Moscow General Assembly in 1958. In particular, Soviet astronomers, who were very active in radioastronomy,[6] helped a group of young Chinese from Peking Observatory, led by Wang Shouguan, to start a Radio Astronomy Section so that could build copies of solar radiometers. Unfortunately, after 1959 relations between the USSR and China started to degrade as the Soviet leader Nikita Khrouchtchev, in the wake of his destalinization policy, which Mao considered as a treason, engaged diplomatic talks with the USA to reduce mutual tensions. As a result, the Sino-Soviet cooperation in astronomy diminished drastically: Since China had withdrawn from the IAU, Chinese astronomers were essentially cut off from the outside world.

This did not deter the ambitious Chinese radioastronomers: Their next project would be a solar imaging radiotelescope. The idea was to start with a copy of an instrument built by the Australian radioastronomer and electrical engineer Wilbur Norman ("Chris") Christiansen. He had pioneered the technique of radio-interferometry, mainly to observe the Sun, in particular building in 1957 near Sydney a double array of 32 radar-like antennas (having a diameter of 6 m) crossing each other at right angles, dubbed the "Chris Cross Telescope." So Wang invited him, on behalf of the Chinese Academy of sciences, to come and visit the group in

---

[3]The Assistant General Secretary, who would oversee the organization of the Prague General Assembly, Luboš Perek (1919–2020), was Czech. See his IAU obituary: https://www.iau.org/news/announcements/detail/ann20034/?lang=

[4]i.e., Cheng Maolan. See Montmerle et al. (2022), and below.

[5]Zhou (2022).

[6]Miley (2022).

Fig. 5.1 Wang Shouguan with W.N. Christiansen with the Fleurs radiotelescope (near Sydney) in the background, 1964. (Credit: *People's Daily*, 10/07/2019, article in Chinese by Shi Fang 施芳, photo Wang Shouguan)

China, which he was able to do in 1963, i.e., after China had withdrawn IAU; and in 1964, returning the invitation, Wang and a colleague would go to Sydney (in spite of the fact that Australia and China did not have diplomatic relations) (Fig. 5.1).[7] As it turned out, W.N. Christiansen was one of the Vice-Presidents of the new Executive Committee in Prague, so he was well acquainted with the Chinese situation.[8]

The construction of the telescope in Shahe (Hebei province, about 300 km south of Peking) made good progress, in spite of the difficulties (the whole technique was new to them), and Christiansen made a second visit in 1965: At that stage, four antennas had been built.

---

[7] The whole story is told in the "recollections" by Wang Shouguan (2009) reproduced in Wang Shouguan (2022), updated in Wang (2016).

[8] W.N. Christiansen would be back on the IAU scene 10 years later, see below.

Fig. 5.2 Cheng Maolan (with glasses) visiting French colleagues at Observatoire de Haute-Provence, Southern France, in spring 1966. The woman to the right is Marie Bloch, from Lyon Observatory, a close collaborator from the time he was living in France. To the left is Mme Anne-Marie Couder, wife of André Couder. Couder was responsible for the optics of the OHP telescopes, and had been IAU Vice-President at the time of the Moscow General Assembly (1952–1958). (Courtesy Mira Véron, OHP)

Another example of a Chinese astronomer traveling from China at that time is Cheng Maolan, Director of the Peking Observatory, having returned to China in 1957 after having spent 32 years in France (and having been part of the Chinese delegation in Moscow), who came back to visit his long-time French colleagues at the Observatoire de Haute-Provence in spring 1966 (Fig. 5.2).[9] Other examples probably exist, but were not recorded in the IAU archives (contrary to Cheng's visit).

However, on May 16, 1966, Mao Zedong launched the "Great Proletarian Cultural Revolution" (无产阶级文化大革命), or "the Cultural Revolution" for short. Fu&Ye write:

> Cultural activities such as education, sciences and others were all put on hold and even were degenerating. Lots of scientists and scholars were put aside away from their works and social ranks. Under this condition, their contacts with the international science community were rather dreams than reality. (...) The Mainland's relation with the IAU also entered into 'the Ice Age'.

---

[9]Montmerle et al. (2022).

For astronomers however, the situation may have been not so harsh as it was for other population categories, although differing from place to place. Wang Shouguan wrote in 2009:[10]

> When the Shahe Experiment was entering its final stage, China began sliding into a 10-year period of chaos, and all work was derailed, and even stopped completely at one stage. Nevertheless, by 1967 a meter wave 'Christiansen array' of 16 east-west elements was installed at the new Miyun Observing Station

[about 50 km North-East of Peking]. Altogether,

> After 1966, work at observatories throughout China stopped for a time, but the situation relaxed somewhat during the 1970s and from time to time we were able to carry out some work at the Miyun Observing Station.

As for Zhang Yuzhe himself:[11]

> [his student] Lu Benkui[12] reported that Zhang Yuzhe was left untouched in the beginning. And they were able to continue to work on this [P651] project,[13] but soon enough Zhang was labeled as one of 'the reactionary (bourgeois) academic authorities' just like the great majority of Chinese intellectuals and scientists at the time. Zhang was politically persecuted by the institutions that he worked for and some of the colleagues that he worked with. He was asked to stop working. (. . .) Based on what we know, if Zhang Yuzhe was partially spared because he was a dedicated science scholar with international importance and name recognition, his wife Tao Qiang [who was a math and geometry teacher] would suffer a terrible fate. (. . .)

(In particular, she was brutally harassed by Red Guards who were her students and others from her school.)

The most violent phase of the cultural revolution lasted about three years (1966–1969), after which the party line (defined at the IXth Congress of the Chinese Communist Party, April 1–24, 1969) became to restore order using the Army, in particular against the Red Guards. However, internal struggles for power lingered on, and the Cultural Revolution is considered to have declined while lasting until Mao's death in 1976.

Not too surprisingly, during this "Ice Age," the successive Executive Committee meetings after #29 in Prague put the "China conflict" on the backburner, reluctantly leaving the Chinese astronomers to their fate until geopolitics could change the situation. In the IAU archives, in spite of the successive changes in the Executive (the Presidents: Pol Swings, 1964–1967, and Otto Heckmann, 1967–1970, and their respective General Secretaries: J.-C. Pecker and L. Perek), there is no correspondence or documents about China—and no news from Taiwan either.

---

[10] Wang Shouguan (2009).

[11] Zhou (2022).

[12] Lu Benkui (born in 1941) would later become the Director of the Purple Mountain Observatory (1996–2000).

[13] Project P651 was the code name for the first Chinese artificial satellite (eventually launched in 1970).

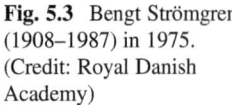

Fig. 5.3 Bengt Strömgren
(1908–1987) in 1975.
(Credit: Royal Danish
Academy)

5.2   The Turning Point: China Admitted to the UN

Then, a fundamental change occurred: Proposed by Albania at the 26th UN General
Assembly, Resolution 2758 to admit China and simultaneously expel Taiwan was
adopted on October 25, 1971, in New York, by 76 votes in favor, 35 against, and
17 abstentions. By the same token, China took Taiwan's seat at the Security
Council.[14] But more importantly for the IAU, it also took Taiwan's seat at
UNESCO.

The IAU Executive responded immediately, by way of its General Secretary, now
C. de Jager, a famous solar astrophysicist (Dutch), elected at the Brighton General
Assembly the year before, along with Bengt Strömgren, a specialist of stellar
atmospheres (Danish), as President (Figs. 5.3 and 5.4).[15] Apparently not knowing
exactly whom to contact in China after 10 years of silence, and likely wondering
what had become of Y.C. Chang and of the Chinese Astronomical Society in the
meantime, de Jager wrote a long letter, dated November 4, 1971, to "The astronomy

---

[14]China had successfully exploded its first atomic bomb in 1964 in the Sin-kiang (Xinjiang, 新疆)
autonomous region (the Uighur region, near Mongolia).

[15]Cornelis ("Kees") de Jager would become ICSU President in 1978.

Fig. 5.4 Kees de Jager
(1921–2021) in 1967.
(Credit: Rob Rutten,
University of Utrecht)

council of the People's Republic of China, Academy of Sciences, Peking, People's
Republic of China":

> In the name and on behalf of the International Astronomical Union, I wish to congratulate
> you most sincerely on the People's Republic of China joining the United Nations Organi-
> zation. I am greatly satisfied and happy that the obstacles which prevented this act during
> 22 years have now, finally, been removed. (. . .) Permit me therefore most cordially to invite
> you to join our ranks. (. . .) I hope very sincerely that this letter of mine may be the beginning
> of a friendly and fruitful cooperation between the astronomers of the People's Republic of
> China and the International Astronomical Union.

Apparently, this letter was either not delivered, or simply put aside, because there
was no answer, since de Jager used the same address, nearly a year later
(on September 5) to send a letter of invitation to the 1973 General Assemblies, in
Sydney or in Warsaw.[16] This time, he received a poorly typed answer, dated

---

[16]In 1973, the General Assembly was scheduled to take place in Australia (the first in the Southern
Hemisphere), but the Polish astronomers came late insisting to hold an "extraordinary" General
Assembly in Poland to celebrate Nicolaus Copernicus' 400th birth anniversary. They also argued
that Sydney was a too costly destination for Eastern astronomers. As a compromise, the Executive
Committee accepted the idea to hold *two* General Assemblies the same year. The Business sessions
and other administrative duties however took place only in Sydney.

December 30, 1972, which, almost to the word, echoed the IAU-China exchanges of 1960. The sender was Hung Szy-yi, Secretary of the Chinese Astronomical Society:

> I regret to inform you that for reasons known to all, the legal status of the Chinese Astronomical Society in the IAU is still being occupied by the 'representatives' of Chiang Kai-shek clique repudiated by the Chinese people, and we are, therefore, unable to attend these activities arranged by the IAU. We believe that, following the development of the friendship between the peoples of China and the world, the friendship and cooperation between the astronomers of China and the world will be strengthened.

But with the admission of China to the UN, and at least from in the IAU archives, the American community (and the IAU General Secretary) clearly took the lead in the reflections about the new situation of the IAU *vis-à-vis* Taiwan. Bart Bok,[17] the American Vice-President, had some exchanges with de Jager, but the other members of the Executive Committee, including Strömgren,[18] apparently did not make significant contributions, at least in writing. Also, Harrison Brown, as Foreign Secretary of the National Academy of Sciences (of the USA),[19] wrote to the Chairman of the US National Committee (USNC) for the IAU, Jesse Greenstein, on November 15, 1971:

> Concerning the possible relationship of the People's Republic of China to the IAU, it should be stressed that nations are not the adhering bodies, rather they are organizations such as academies which represent the scientific communities in particular geographic areas. It seems to me that the Academy of Sciences in Peking is going to have to recognize that the Academia Sinica in Taiwan can be a member of the Unions and of ICSU itself for as long as the two areas are separated politically. From ICSU's point of view, the situation is very similar to the situations with respect to North Korea versus South Korea, North Vietnam versus South Vietnam, and East Germany versus West Germany, all of whom belong to ICSU.

On this last point, however, Brown makes the usual error: Whereas there were at the time really, politically speaking, "two Koreas," "two Vietnams" (unifying four years later after a terrible war), and "two Germanies" (until the peaceful reunification of 1990), China's mantra was (and still is) that there is only one China, not two—which makes a fundamental difference.

In turn, in passing the baton of Chairman of the USNC to D.S. Heeschen, J. Greenstein wrote on November 22:

> We must take, as scientists, a position that science is beyond politics, and that scientific representatives from quite adverse political worlds can live together in the IAU. (...) If the Mainland Chinese suggest a visit of a delegation of American scientists... *I offer myself as a lamb for the sacrifice.*

---

[17] 1906–1983. More below.

[18] In a private email exchange between C. de Jager (then aged 99!) and myself, on January 11, 2020, to a question I had about Strömgren, de Jager replied: "He hated unnecessary correspondence... he did not like to answer letters."

[19] Harrison Brown (1917–1986) was an American physicist and nuclear chemist. He would be ICSU President in 1974 (see Sect. 6.2).

But de Jager was much more lucid. In a letter to B. Bok (December 7), de Jager expresses his analysis, which would be the backbone of the future negotiations (and as such is quoted here rather extensively; the italics are mine):

I think that our problem is a difficult one, with many subtle points. One of these may be *the name under which Taiwan remains a member of the Union.* As you know, the IAU, like ICSU, is of the opinion that any region on earth [sic] where science is practiced at a reasonable standard must be in a position to be represented in the international scientific bodies. On the basis of that principle, our Union accepted the German Democratic Republic [in 1962] even at the time that it was recognized by a very few countries only, and on the basis of the same principle we do not want to expel South Africa, whatever one may think of its Government.[20] As Harrison Brown rightly remarks, the Nations are not the adhering bodies.

(...)

The People's Republic will certainly ask us to expel Taiwan from the IAU as a condition for its own re-entry. Personally, I would never accept such a condition, and although I do not believe that much good astronomy is practiced in Taiwan, I would rather leave my post as General Secretary if the Union would decide to expel Taiwan for Mainland China.

(...)

Life might become easier for us when Taiwan would henceforth accept to be listed just as Taiwan, deleting 'Republic of China'.

The next major geopolitical development, for the IAU, and in particular for astronomers involved in pursuing or developing contacts with their Chinese colleagues, was undoubtedly the visit to China by the American President Nixon, from February 21 to 28, 1972: The handshake between Mao and Nixon instantly made the news around the world. The trip had been planned long in advance, in fact even before China was admitted to the UN, but in spite of this apparent blow to American diplomacy, Nixon decided to abide by his decision to meet Mao Zedong and especially his Prime Minister Zhou Enlai, and explore ways to normalize relations between the two countries. Of course, this historic visit did not solve all the problems, especially the central one, the status of Taiwan: The USA would not break off their diplomatic relations with Taiwan until January 1979—another key date for the IAU–China–Taiwan relations. But as of Nixon's visit, permanent contacts were officially established between China and the USA, which in turn triggered a wave of initiatives on the part of the scientific community worldwide, and in particular of ICSU, as well as, of course, the IAU.

But there would be an unexpected obstacle: UNESCO.

During the Executive Committee meeting (#36) in Copenhagen, held on August 28 to September 1, 1972, and presided by Strömgren, the subject of "Mainland China" was discussed briefly. First, it was acknowledged that the letter of invitation by the General Secretary (C. de Jager; on November 4, 1971) did not receive any answer (see above), so that "good services of Professor Christiansen" and other astronomers should be encouraged. Also, addressing the point raised by de Jager in his letter to Bok on December 7, 1971 (see also above), the Executive Committee

---

[20] At the time, South Africa was under the Apartheid regime. The talks to suppress this regime started in 1974, but they lasted for almost twenty years until the 1992 referendum.

declared that *"the expulsion of Taiwan from the Union as condition for the Chinese [People's] Republic joining it would be intolerable."*

This was exactly the discussion that took place in Peking on November 8, 1972, between the President and the Executive Secretary of ICSU (resp. J. Coulomb and F. A. Stafleu, elected at its 14th General Assembly in Helsinki the preceding September),[21] and Prof. Chang Wei, Vice Chairman of the Revolutionary Committee of Tsinghua University (Peking), and member of the Executive Board of UNESCO,

> in order to begin informal discussions concerning the possibility of the Academy of Sciences of the People's Republic of China adhering to ICSU". (. . .) "The President explained that ICSU, as a non-governmental organization, had independent groups of scientists as adherents and not governments: The acceptance of a group of scientists did not imply recognition of the legal status of the government of the geographical areas concerned." (. . .) "The President . . . put forward a suggestion that in the first instance the Academy in Pekin [sic] might consider the possibility of adhering to those Unions which did not have Taiwan as a member. The proposal was met with interest. Professor Chang explained that . . . he would discuss the situation with the delegation of the People's Republic of China to the UNESCO General Conference. . .

On November 10, Stafleu launched an all-out inquiry to the "Secretaries General of International Scientific Unions" and "Secretaries of Scientific and Special Committees," asking for information about their relations with the Academy of Sciences of the People's Republic of China, more precisely (*1*) a list of their national adhering organizations, (*2*) an indication of their contacts with: (*a*) the "Academy of Sciences, Pekin," (*b*) the "Academy of Sciences, Taipei"; and (more importantly) (*3*) "if you would welcome any initiative leading to an application for admission from the Academy of Sciences, Pekin," inviting however the Secretaries to refrain from any direct contact with the Academy of Sciences, Pekin, until further discussions were to be held on November 23.

The answer from the IAU is not in the archives, but the one to item (*3*), in particular, is clear, since de Jager had already tried to contact the Chinese Academy of Sciences after the admission of China to the UN—getting the "standard" negative answer only over a year later, on December 30.

5.3   A Man with a Mission: The Goldberg Presidency

The next year (1973) would see an acceleration of events, from many sides: UNESCO, ICSU, IAU, internal politics in China. It was a crucial year, where solutions began to emerge, albeit laboriously.

---

[21] Jean Coulomb was succeeding V.A. Ambartsumian, elected in 1968; Frans Antonie Stafleu was in his second term. See below, Sect. 5.6.

On the UNESCO front, at first the news was good. On February 2, 1973, F. Stafleu, the ICSU Secretary General, answered a letter sent on December 22, 1972, by UNESCO's Director General, René Maheu, concerning "the attitude of the People's Republic of China to non-governmental organizations": *"I am pleased to learn that the representative of the People's Republic of China agreed, in principle, to the granting of UNESCO subventions to certain non-governmental organizations."* Referring to Taiwan and the ongoing discussions with the Chinese Academy of Sciences, and also evoking the pledges of UNESCO, Stafleu continued: *"You may rest assured that ICSU, with its stated policy of freedom for all scientists to adhere to or associate with international scientific activity regardless of race, religion, political philosophy, ethnic origin, citizenship, language or sex, is doing its utmost to resolve this difficult situation."*

On their side, astronomers were trying to reach out to colleagues in China, inviting them (depending on their field of interest) to IAU scientific meetings, with the idea that they would maximize their contacts, or to collaborate on specific projects. But obviously, involving the IAU was not a good strategy: For instance, invited to participate in an IAU Symposium in Poland by the American planetary astronomer Tobias Owen, Y.C. Chang (now back in office as Director of the Purple Mountain Observatory and President of the Chinese Astronomical Society) replied in a thankful letter dated January 15: *"I regret to tell you that for obvious reasons, we have withdrawn from membership of the IAU since 1959, and on principle we shall not participate in any activity sponsored by the IAU."* Owen sent a copy of Chang's letter for information to Strömgren, so that the IAU Executive Committee would be aware of the (unchanged) China policy, thus confirming the negative reply sent to de Jager a couple of weeks before by the Secretary of the Chinese Astronomical Society.

Nonetheless, Leo Goldberg tried once more to convince Y.C. Chang to attend the Sydney General Assembly: On July 25, less than a month before the opening, he wrote: *"The exclusion of such a large area of the earth as China from participation in international projects seriously handicaps the progress of astronomy,"* to which Chang replied on August 17, in a friendly but slightly irritated tone: *"Now is time to put an end to this absurd situation. Our fundamental principle is: So long as the members of Chiang's Clique have not been expelled from IAU, we could not consider taking part in any of its activities."*

In the end, seven Chinese astronomers went to Sydney and attended the General Assembly, but they were all from Taiwan.

In his address to the General Assembly as incoming IAU President, Goldberg (Fig. 5.5) concluded:

Finally, although the Executive Committee faces many important and difficult tasks before the Sixteenth General Assembly meets in Grenoble, none is more urgent than finding some way to effect the return of our colleagues from China to the IAU. On this hopeful note, I will conclude and look forward to seeing many of you in Warsaw in a few days,[22] and all of you in France in 1976.

---

[22]To attend the Extraordinary General Assembly (see Footnote 16).

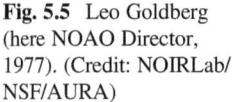

He would repeat his statement on October 3 in a "private" circular letter to the new Executive Committee: *"I came away from Sydney with the fairly strong feeling that the membership of the IAU will be unhappy if we have not succeeded in restoring China to membership by the time of the next General Assembly."*

Contrary to Strömgren, and as we have already noticed during the preparation of the Berkeley General Assembly in 1961, and (to a lesser extent) during his six-year term as Vice-President thereafter, Goldberg loved to write letters, and thus the IAU archives contains an extraordinary record of his presidency.

At the time of the Sydney General Assembly, Goldberg was at the height of his career.[23] Born in New York City on January 26, 1913, to Jewish immigrant parents of Polish origin having fled Tsarist Russia, and gifted for maths and science, he had graduated in astronomy from Harvard University (PhD in 1938). He was then appointed to the McMath-Hulbert Observatory in Michigan in 1941, where he studied solar physics and spectroscopy, for which he became well known (solar flares in particular) and also worked on a wartime anti-submarine project (see Footnote 11 in Chap. 3). He became Professor at Harvard University in 1960, and Director of the Harvard Observatory in 1966. He was very influential in the development of solar physics and spectroscopy from space and eventually became

---

[23] For details, see his NAS Obituary by Aller (1997).

Fig. 5.6 George
Contopoulos in Sydney,
1973. (TM is extremely
grateful to G. Contopoulos
for having kindly sent me
this picture from his private
collection.)

Director of the National Optical Astronomy Observatory (NOAO) in Tucson,
Arizona, in 1971, his position when he became IAU President.

The new General Secretary was George Contopoulos, a renown Greek specialist
of celestial mechanics and dynamical systems (Fig. 5.6),[24] with Edith A. Müller as
Assistant General Secretary (from Geneva Observatory, she was the first female
astronomer elected to this position; she would play an important role as General
Secretary of the next IAU President, Adriaan Blaauw).[25] The new Vice-Presidents
were J. G. Bolton (Australia), Ch. Fehrenbach (France), and W. Ivanovska (Poland),
while the second-term Vice-Presidents were B. Bok (USA), B.Lovell (UK), and
E. Mustel (USSR).[26] The Advisors were the past President and General Secretary,
resp. B. Strömgren and C. de Jager, whom we met previously. The (im)balance

---

[24] George Contopoulos (Κοντόπουλος, or Kontopoulos in modern Greek), born in 1924.

[25] Chapter 7.

[26] *First term:* John Gatenby Bolton (1922–1993); Charles Fehrenbach (1914–2008), Director of the
Haute-Provence Observatory in Southern France, established in 1937 and built progressively during
the war (Fehrenbach, 1990), see also Montmerle, Zhou and Gomas, 2022; in the IAU archives, the
letters Fehrenbach exchanged with L. Goldberg were all in French); Wilhelmina Iwanowska
(1905–1999). *Second term:* Bartholomeus Jan "Bart" Bok, Dutch born (1906–1983); Sir Bernard
Lovell (1913–2012, founder of the Jodrell Bank radiotelescope), Evald R. Mustel (1911–1988).
Due to health problems, Bok was replaced the following year by Per Olof Lindblad (the son of
Bertil Lindblad), from Sweden (1927–) for one term only (1974–1976). P.O. Lindblad would be
Vice-President again later (1985–1991).

between East and West was not as bad as it looks, since Bok had been genuinely concerned about China during his first term (as was of course Goldberg), and Lovell had a long history of collaboration with Soviet astronomers (with the Jodrell Bank Mark I radiotelescope).[27] And, following a certain French tradition dating back to before WWII (see above the excerpts from Fu&Ye), Fehrenbach was also in contact with Chinese astronomers.[28] All in all, a reasonably well-prepared Executive Committee to take an active role in the new developments of the "China problem."

In his above-mentioned circular to the new Executive Committee, Goldberg makes further remarks that would, to some extent, serve as guidelines for further discussions on China: (1) stay in touch with ICSU in their first efforts to discuss with China, and (2) involve USA-based Chinese scientists having contacts in China to gather information about their perception of possible relations with the IAU, even unofficial ones. Goldberg writes:

> As you know, a small committee of ICSU officers is about to leave for discussions on China, and I suppose that if a satisfactory formula is found that will restore China's membership in ICSU, our own problems will be solved as a matter of course. On the other hand, if the efforts by ICSU should fail, I believe we should continue to try to find an independent solution.

Followed by:

> I spoke on the telephone yesterday with Prof. C.N. Yang, the Stony Brook physicist who has made several visits to China, most recently last summer, and has close connections with highly placed scientists and politicians in China.[29] It is his belief that the Chinese would not object to the presence at IAU meetings of astronomers from Taiwan, provided that there is no indication that IAU regards Taiwan as a separate country independent of China.

In his reply, on October 11, de Jager wishes to rectify Goldberg's circular and update him on the situation:

> The Academy in Taipeh [sic] no longer claims to represent China, but is very careful in the last year to claim representation of Taiwan only. This makes the situation fairly analog to that of the two Germanies and the two Koreas,[30] with the exception that Peking does claim to represent the whole of China,

adding *"It would not be advisable for ICSU or for IAU to exclude from its membership a politically independent geographical region where science is practiced."*

The new, "low-profile" step by the Academia Sinica in Taipei was thus very significant. Of course, this change resulted from the expulsion of Taiwan from the UN, a "country" (in the IAU definition) still de facto independent from the people's Republic of China but, in spite of its population having grown to about 15 million, banned overnight from the official list of nations.

---

[27] Now the "Lovell Telescope." See Miley (2022).

[28] Montmerle et al. (2022).

[29] Chen-Ning Yang (杨振宁), and his fellow particle physicist Tsung-Dao Lee (李政道), were awarded the Nobel Prize in 1957.

[30] S. Korea had just adhered to the IAU, at the Sydney General Assembly.

The following day, Goldberg tries again to convince Y.C. Chang, mentioning the ICSU efforts:

> ... My position is that there is indeed but one China, namely, the People's Republic of China. (...) Taiwan's membership in the IAU does not deny the right of the Astronomical Society of the People's Republic of China to represent the astronomers in Taiwan. It simply acknowledges the practical situation that the Chinese astronomers are not, in your own words 'in a position to keep contact with the astronomical workers in Taiwan.' This was already understood at the time of Taiwan's adherence in 1959 when it was expressly stated by the IAU that the Astronomical Society in Taipei would represent only astronomers in the geographical area of Taiwan." (...) "I have learned that the officers of ICSU are now visiting your country and undoubtedly they are seeking to explore with the Academia Sinica in Peking some formula by which China could reassume membership in ICSU without at the same time depriving scientific workers on Taiwan of membership in the international scientific community.

Goldberg logically concludes: *"therefore it seems best to me to postpone consideration of our own problem until our discussion between ICSU and the Academy in Peking have been completed."*

This time, in an unusual exercise, several Vice-Presidents independently answered Goldberg's circular. Fehrenbach (on October 15; in French, here translated): *"I had the possibility of talking to Prof. Coulomb, ICSU President, before he left for China. His impression is that the Chinese will never adhere to the Unions that have representatives from Taiwan. It is obvious that Taiwan's membership in no way represents China, and under these conditions I would vote against its membership and I would find justified that Academia Sinica files a request to represent China. The Academy in Taipeh [sic] can only represent Taiwan. (...) In any case, it is wise to wait for the results of the ICSU mission before undertaking any initiative."* In other words, Fehrenbach seemed in favor of expelling Taiwan as the price to pay for asking China to rejoin the IAU. On the other hand, Iwanowska (on October 25) thought that *"the Academia Sinica in Peking ... will be the proper body to represent astronomers living on Taiwan as soon as this territory will be under the control of the Chinese Popular Republik [sic]. Up to this time they probably can be represented by the Academy of Taipeh which could apply to represent the territory of Taiwan."* So, perhaps unexpectedly since she was from the "East," Iwanowska fully supported the IAU's justification of its 1959 decision, as recalled in Goldberg's October 12 letter to Y.C. Chang. In contrast, Bolton (from the "West") adopted a radically opposed position (which will surface again later), in a handwritten letter dated October 22: *"I have discussed this matter with W.N. Christiansen* [past Vice-President; see above[31]] *(...) it is quite clear that an error was made in the original admission of Taiwan as a member of the IAU. I think that the present executive should admit this as a preliminary to any discussion with China* [underlined in the letter] *(...)."* Much later (on December 24), Mustel (second-term Vice-President) gives a more "politically correct" answer: *"The position of the Soviet representatives*

---

[31] W.N. Christiansen was known to have long time collaborations with Chinese radioastronomers (e.g., Wang Shouguan, 2022).

[32] in the Executive Committee of the IAU on this problem was always quite straightforward. We are sure that the Chinese Republic should be considered as the only full right representative of the Chinese people. (...) The Executive Committee [should not] *involve additional complications and* [should] *admit the Chinese People's Republic as the legitimate representative of the nation with all logical successions."* Adding that *"Taiwan cannot represent the astronomy of China."* However, Mustel did not explicitly ask to expel Taiwan.

This exchange of views is important on several grounds: (1) All of the freshly elected Vice-Presidents replied to the President; (2) as individuals, they candidly expressed opinions that, seen from afar, could be taken as opposite to the political views of their countries; (3) those with hands-on experience in Chinese collaborations (Bolton, as the voice of Christiansen; Fehrenbach) are clear supporters of China at the expense of Taiwan. This illustrates why one should refrain from making a simplistic one-to-one correspondence between country citizenship and Executive Committee membership, when issues are raised that have political implications such as these.

However, the exchange also shows a certain lack of knowledge of how the IAU works, and in particular (again!) of its Statutes. On reacting to the replies of the Executive Committee on November 12, Goldberg makes a legal point: *"As long as Taiwan does not claim to represent the astronomers on the Mainland of China, there is no legal way in which they can be expelled from the IAU as China demands. Thus* [essentially paraphrasing his letter above to Y.C. Chang about ICSU] *we must continue our efforts to find a mechanism by which China could return to the IAU without at the same time depriving astronomers on Taiwan of membership in the international astronomical community."*

Then, Goldberg *"urges that all of us make every effort to invite individual Chinese astronomers to attend international meetings and to participate in cooperative projects. (...) If you can identify one or more active astronomers in China, please communicate their names to George Contopoulos."*

5.4 The ICSU Mission to China

Meanwhile, the "ICSU Mission," as it called itself, traveled to China "to exchange views on the establishment of relations between Chinese scientists and ICSU." The Mission consisted of three scientists: Jean Coulomb (President) (Fig. 5.7), F.A. Stafleu (Secretary General) (Fig. 5.8), already mentioned, and also F.W.G. "Mike" Baker (Executive Secretary), a botanist and climate specialist. The inviting body was

---

[32] It is very unusual (and inappropriate) to consider the IAU Vice-Presidents as "representatives" of their country, i.e., having official mandates, for instance in case of a vote. This is the prerogative of National representatives only. Even Kukarkin, when challenging the Executive Committee majority to admit Taiwan in 1959, and again when he submitted his resolution to the Berkeley General Assembly in 1961, never implied some form of national mandate.

Fig. 5.7 *(Left)* Jean Coulomb (1904–1999) was a French geophysicist. He had a bright career in research management and became President of IUGG (1967–1971), then of ICSU in 1972. He was succeeded by Harrison Brown at the ICSU 16th General Assembly of 1974 in Istanbul (see also Appendix B.3). (License: CC BY-SA 4.0: https://commons. wikimedia.org/wiki/File: Jean_Coulomb.jpg)

now clearly identified: For the first time, instead of the name "Academia Sinica Peking," or "Peking Academy," easily misleading (and probably wrong, for lack of better information) when referring to "Academia Sinica Taipei," or "Academy in Taiwan" at the same time, the invitation came from the *"Scientific and Technical Association of the People's Republic of China."* The Mission visited "a number of Research and Teaching Institutes" (which are not specified). The contact person was Prof. Chang Wei, the Deputy representative of China at the UNESCO General Conference, whom Coulomb and Stafleu had met a year before in Peking (see above). On the Chinese side, in addition to Prof. Chang Wei, the discussion involved Prof. Chou Pei-yuan (周培源, Zhou Peiyuan), Vice Chairman of the Scientific and Technical Association, and Vice-President of the Revolutionary Committee of the University of Pekin [sic], and other (unnamed) colleagues.

The five-page Report of the Mission, dated November 5, 1973, summarizes the discussions that took place a few days before, on October 18. The preamble warns:

The present document is intended to provide members of ICSU with information on the discussions. It does not in any way commit our Chinese colleagues. It was agreed at the beginning of the discussions that they were exploratory and not a negotiation, so as to allow both points of view to be presented frankly in a spirit of mutual comprehension. (...) We hope that we have recorded faithfully our host's thoughts. I present here the highlights of this

Fig. 5.8 *(Right)* Frans
Antonie Stafleu
(1921–1997) was a Dutch
botanist, who had chaired
the Institute of Systematic
Botany at the University of
Utrecht. (source: http://
www.biografischportaal.nl/
en/persoon/36317737, in
Dutch)

important report, because most of the arguments it contains would still be valid by simply
replacing "ICSU" by "IAU."

As expected, the Chinese scientists opened the discussion by putting forward the
new status of China after its admission to the UN in 1971, and the resulting
Resolution of the UNESCO Executive Board meeting of October 8, 1973 (to be
presented in more detail later, Chap. 6). This Resolution stipulated, in essence, that
UNESCO would follow the UN decision, i.e., systematically replace Taiwan by
China in all bodies and associations having a link with it, ICSU being one of them,
and asking ICSU to enforce the Resolution and instruct its Unions and Associate
members to do the same without delay.

Coulomb then recalled the main characteristics of ICSU:

(i) it is an organization which is completely independent of UNESCO; (ii) National Members of ICSU represent groups of scientists from the same region with no account taken of
the political status or form of government of the region concerned. ICSU had welcomed,
independently of any UN resolution, the scientists of the two parts of Germany, of Korea,
and Vietnam. The East African Academy represents scientists from three different countries.

Replying in turn, Prof. Chou Pei-yuan (周培源, Zhou Peiyuan in pinyin)

considers that ICSU is, to a great extent, the mother of the Unions and of the Committees. *He
considers it impossible to negotiate with the children without first reaching an agreement
with the mother.* He believes that it is not possible for a scientist to divorce himself from his
country of origin. The presence, regardless of their label, of scientists associated with the
government of Tchang Kai-chek, would signify the ICSU accepts the existence of an

independent Taiwan, and this the people of China cannot accept. The historical and effective participation of China in various Unions (e.g., IUPAP, IAU) had been destroyed by the imperialist introduction of Taiwan as a deliberate political act.

At this point, Baker drew Prof. Chou Pei-yuan's attention to

the sorry experience of the International Research Council which had introduced politics at the foundation in 1919 when only Allied and Neutral Powers had been allowed to adhere. It took 12 years to rectify this situation and since then ICSU had no wish to meddle again in politics.[33]

Then,

Professor Chang Wei noted with appreciation the desires of Professor Stafleu (who 'expressed the hope that Chinese scientists would be able to participate in the international scientific activities organized by ICSU') and Mr. Baker but stated that there are two main difficulties:

1. The idea of a 'National' Member;
2. The fact that ICSU seems to consider Taiwan as a nation.

Coulomb then explained that ICSU was governed by its Statutes and by its members during General Assemblies so that any modification of the status of Taiwan could not be made without their approval. Stafleu pointed out that a change in the Statutes *"required that the Assembly decision has a 2/3 majority vote."* He agreed with Prof. Chang Wei that *"it was necessary to proceed in a positive fashion with prudence."*

In spite of the courteousness of the discussion, its conclusion for ICSU was the same as it had been for the IAU: deadlock.

5.5 Astronomers in China: Act II

In the fall of 1973, not only were there high-level official discussions going on between ICSU and Chinese scientists in Peking, but contacts were resuming between Chinese and Western astronomers. We have already mentioned the early exchanges during the 60s in the radioastronomy field between the Australian astronomer W.N. Christiansen and the Peking group of solar radioastronomy under Wang Shouguan, and between Fehrenbach and Cheng Maolan and his students. Ten years later, as the Cultural Revolution was slowly ebbing away after the Xth Congress of the Chinese Communist Party (held in August 1973), more astronomers began to visit China. As recorded in the IAU Archives, this included J.A. Hogböm, from Sweden, and G. Miley, then a young radioastronomer from Leiden, the Netherlands, and more exchanges would develop over the years. But many other collaborations may have developed elsewhere.

---

[33] E.g., Blauw's book, Chap. 3, for a summary. See also Fauque and Fox (2022).

Fig. 5.9 George Miley (with Hawaiian shirt), Y.C. Chang (with cap) and other Chinese astronomers at the Purple Mountain Observatory, 1973. (Photo G. Miley, digitized original courtesy of *Sky & Telescope*)

Miley wrote a brief report in October, with details about his trip to the three main observatories in China (Nanking, Peking and Shanghai), and to the solar radioastronomy station of the Mi-yun reservoir, housing an interferometer started with the help of W.N. Christiansen and now completed.[34] Interestingly, Miley reports that

> It is difficult to assess the impact of the Cultural Revolution on Chinese astronomy. Of course, the universities have gone through a period of unprecedented turbulence. Virtually closed for six years, they are now gradually re-admitting students under a drastically reformed system. (...) By contrast with the universities, the observatories and research institutes have emerged relatively unscathed from the Cultural Revolution.

New instruments appeared to be under construction, in particular a 2-m optical telescope to be built in the Yen-Shan mountains, 160 km to the NE of Peking. Miley's trip gained widespread visibility with an article published in the American popularizing journal *"Sky & Telescope"* (Miley, 1974) (Figs. 5.9 and 5.10).

With the "renaissance" of the Chinese astronomy after the Cultural Revolution, as witnessed by these astronomers (acting independently), the issue of the return of

---

[34] Miley (2022); Wang Shouguan (2022), and Shouguan Wang (2006—following the Western order of names).

Fig. 5.10 A group of radioastronomers from Peking Observatory (now the Headquarters of the National Astronomical Observatories of Chinese Academy of Sciences, Beijing), in front of one of the 6-m dishes built as part of the Miyun station interferometer in 1973. The head of the group was Wang Shouguan (center). This image was first published in Sky & Telescope (March 1974 issue, p.152), in an article entitled "Astronomy in China", by George Miley, from Leiden Observatory in the Netherlands. See Sect. 5.5 (Photo G. Miley; digitized original courtesy of *Sky & Telescope*; with permission from Wang Ying, Wang Shouguan's daughter)

China into the IAU became, to some extent, progressively more pressing scientifically than politically, especially since that meanwhile Taiwan, still under the dictatorship of Chiang Kai-shek, seemed to develop, albeit very slowly.

5.6   A Hopeless Situation

Unfortunately, overall the year 1974 did not bring much progress. The issue of Taiwan vs. China was going in circles, with much the same rhetoric on both sides, and even some misunderstandings about the respective actions of ICSU and the IAU.

For instance, on April 24, answering a letter from Lovell (which is not in the archives), "as means of bringing the entire Executive Committee up to date on the situation as I now see it" Goldberg wrote:

> The visit by the ICSU delegation to China last fall was not very successful and events since that time have convinced me that we probably cannot look for ICSU for leadership in solving the Chinese problem. (...) it seems unreal to go on pretending that Taiwan is a separate country. (...) We do have to face the fact that at present China is physically unable to represent Taiwan in the IAU and therefore separate representation is necessary, at least for the time being.

C. de Jager, still Advisor to the Executive Committee, and now President of COSPAR,[35] wrote to Goldberg (May 7), saying that, in his letter to Lovell, he (Goldberg) did not appreciate the action of ICSU correctly because he had not taken into account Taiwan's new position:

> I think your information [ICSU on China] is incomplete and therefore your judgement not wholly correct (...) Why is it that we do not progress? There are two clear reasons. First, the National Representative of Taiwan in ICSU, and the Taiwan Government, do not want Taiwan to be called a province of the People's Republic of China. They even refuse to call their Academy 'Academy of Taiwan' but stick to the ridiculous 'Academia Sinica'. Secondly, China would only accept membership of any ISCU organization if Taiwan would be expelled from all ISCU organizations; also China would not accept a separate Taiwanese delegation in the composition of which they would have no say, even not of that delegation would be called 'Delegate of the Province of China'.

Then, de Jager adds, in a desperate tone: *"So, if you see a solution to this problem, please go ahead."*

On May 20, Goldberg reacted somewhat nervously: *I see no reason for changing my assessment that ICSU has not made any significant progress towards the solution of the Chinese problem (...)* [The IAU Executive Committee]

> will not be satisfied with reliance upon ICSU unless the end result is something more than reaffirmation of the status quo. (...) I hope we can find a solution that is both 'tactful and diplomatic' [this refers to words in the letter by de Jager] but it is even more important that it be rational and reasonable. *After all, Taiwan's original application to the IAU was not exactly motivated by tact and diplomacy and its admission was voted over the bitter protests of two members of the Executive Committee.* It is also interesting to note that China did not withdraw from the IAU until Taiwan became a member, although up until that time Taiwan, but not China, was a member of ICSU."[36]

Since Taiwan had already agreed to drop its claims to represent the whole of China in scientific unions, Goldberg continued to think that an appropriate change in designation could allow progress to be made. The term "Province of China" not being accepted by Taiwan, the suggestion was to drop the political connotation of "Province" and use "part of China" instead. This may well seem today like splitting hairs, but the idea made its way to the ICSU General Assembly that would convene in Istanbul the following September. The resolution was first submitted to the IAU Executive Committee meeting (#39) in the Haute-Provence Observatory (at the invitation of Fehrenbach) on August 27–29, 1974, with J. Coulomb, as (French) ICSU President, attending. After discussion and the approval of Coulomb, the

[35]The *Committee On SPAce Research*, one of the ICSU Scientific Committees, established at its 1958 General Assembly in Washington, in the wake of the launches of the *Sputniks* by the USSR, and of *Explorer-1* by the USA. This General Assembly was the one in which Taiwan would apply to several Unions including the IAU (see above, Chap. 4).

[36]This is a factual error on the part of Goldberg: As noted above (Chap. 4), before 1949 ISCU referred to "Nanking," and after 1949, it referred to "Peking," not to "Taipei." So in fact China had been a member of ICSU since 1937.

resolution to be submitted to the ICSU General Assembly mentioned in particular the suggestion that *"Taiwan be, for the purposes of ICSU, henceforward listed as a part of China, and that this be expressed in all official documents of ICSU by having the designation 'Taiwan' followed by the words '(part of China)'."* The resolution was approved by the Executive Committee, but W. Iwanowska and E.R. Mustel abstained from the voting.

In parallel with the official positions taken by ICSU and the IAU (and other Unions) not to expel Taiwan because their Statutes did not allow it, there were signs that individual initiatives were positively encouraged by Chinese scientists. A case in point is the scientific invitation extended by the Director of the Peking Observatory to Donald Menzel (via the Chinese Academy of Sciences), a former Director of the Harvard Observatory, to discuss in particular issues of planetary and solar system nomenclature. On his way back, on September 26, Menzel sent a brief report from Hong Kong to Goldberg:

> I learned I was the first American astronomer to receive an official invitation to visit China.
>
> (...)
>
> One of my objectives was the obtaining of Chinese names for our lunar nomenclature project. The various scientists, including Prof. Wu Yu-hsun[37] Vice-President of the Academy of Sciences, were reluctant to cooperate because this could be interpreted as approving the IAU and its policies, in particular membership of Taiwan instead of the People's Republic.
>
> (...)
>
> Again and again I heard the statement: 'If U.N. can dismiss Taiwan, why can't the IAU do the same?' It is quite clear that they will never come in as long as Taiwan remains a member.
>
> (...)
>
> On the whole my impressions of China are very favorable. Their industrialization is surprisingly advanced and their stores are full of excellent consumer products in great variety. They were generous hosts and we met the leading astronomers, young and old.

Based on that information, Goldberg decided to send the Vice-President of the Academy of Sciences a long explanatory letter (October 8) summarizing "[his] views on Taiwan" and the history of the relations between China and the IAU, in the light of the issue mentioned by Menzel, i.e., the admission of China to the UN:

> In 1959, when Taiwan was accepted as a member of the IAU, the Executive Committee passed a resolution specifying that it would be representing astronomy *only on the island of Taiwan*. Thus, China's seat in the IAU is not being filled by anyone at the present time—to the great sorrow of all astronomers, it is empty. (...)
>
> 'If U.N. can dismiss Taiwan, why can't the IAU do the same?' The answer is simple: The IAU believes that all astronomers have the right to be represented in the organization, regardless of political considerations. The UN chooses its membership on political grounds; the IAU does not. I need only to remind you that the People's Republic of China was welcomed into the IAU fifteen years before its acceptance by the UN. (...)
>
> If you believe that the only possible solution is the ejection of Taiwan from membership in the IAU, then we have no choice but to abandon temporarily our attempts at a solution and to await political events that would automatically solve the problem. Cancelation of the membership of Taiwan would be illegal according to our statutes, and we could not accept a

---

[37]Prof. Wu Yu-hsun (吴有训, Wu Youxun).

precedent that might make it possible in the future for other countries to be ejected on
political grounds. (. . .)

   It is my view that the position of the Chinese government would not be compromised if,
in the list of members of the IAU, your country should appear as "China" with adhering
organization "Academia Sinica, Peking," while the membership of Taiwan would be stated
as "Taiwan (part of China)," with adhering organization "Academy of Sciences, Taipei". ...
If there were some indication that China would accept this proposal or some variant thereof,
there is, in my opinion, a very good chance that it would be adopted by the IAU.

Unfortunately, there was no answer to this letter on the part of Prof. Wu Yu-hsun
or of the Academy of Sciences. On the contrary, the situation had actually worsened.
In Istanbul, the ICSU General Assembly passed on September 25 a new resolution
on the "Free Circulation of Scientists" (since the first one of the Washington General
Assembly in 1958 already mentioned, there had been three subsequent versions,
issued in 1963, 1966, and 1972; see Greenaway, in particular Chap. 8), and the
invitation to China to become a member of ICSU was renewed (including the IAU
resolution on Taiwan). In reply, the Scientific and Technical Association of the
People's Republic of China issued on November 1 a "Statement" expressing *"the
[Association] and all Chinese scientific workers firm opposition to this resolution,"*
blaming ICSU for not having followed the decision of UNESCO to break off all
relations with Taiwan (using the same wording as the Executive Board resolution
quoted above), and reiterating that *"this is, in reality, creating "two Chinas" or
"One China, one Taiwan."* The statement continues: *"The leading officials try to
justify their action by pleading 'universality' and 'no involvement in politics'. This is
untenable. The fact is that their absurd decision precisely runs counter to the
principles they advertise."* (The full text, in English and in Chinese, is reproduced
in Appendix B.3.)

For Goldberg, that was the answer he had feared (and which he did receive
separately) to his letter of October 8, using the same rhetoric as 15 years earlier. On
December 9, 1974, writing to Contopoulos, he throws the gauntlet: *"I am about
ready now to let the whole matter drop until there is a change in the whole political
situation."*

At that point, it is worth quoting again Fu&Ye:

This [period] was as if a clear blue sky was spotted by dark clouds, a sky that lacking its blue
color was no longer perfect. This exceedingly twisted reality of the international astronomical relations [between China and the IAU] did stir up some discontent among the righteous
astronomers, but it still seemed to lack the wind to blow the dark clouds away.

5.7 Declarations and Unrest at the Grenoble General Assembly (1976)

Apart from the discussions between UNESCO and ICSU, and between ICSU and its
Unions (including the IAU; see Chap. 6) about the status of Taiwan after its
expulsion from the UN, nothing much happened in 1975. Worthy of note, however,
is the record in the Archives of several "private" visits of astronomers to China: for
instance, of P.O. Lindblad from Stockholm in September, meeting in particular

Wang Shouguan, or exchanges with D. Menzel about his trip in 1974 (see above), just to revive the flame. In 1976, Goldberg (March 23) and Wang (July 21), respectively, replied to a report sent by Lindblad, showing some bitterness:

> Unfortunately, Professor Wang seems to have contributed nothing new that might provide a clue as to how we should proceed to solve the China problem. (...)
> I do not think particularly helpful to reiterate that Taiwan was pressured into applying for IAU membership by the US Department of State in order to get the People's Republic of China out. I believe personally that the story is true but there was nothing illegal about the proceedings and there was no way in which the IAU could have refused admission to a qualified country without violating its statutes. I will be very surprised if you hear further from Dr. Wang even though I hope my prediction is wrong.

In fact, Wang Shouguan did reply to Lindblad (this was just one month before the Grenoble General Assembly):

> It is our belief that the visits by you and Dr. Hogböm will promote mutual understanding and friendship between the astronomical workers of our two countries. (...) the present and future leaders of the IAU have the very responsibility to correct the historical mistake of the IAU. (...) In accordance with the 5th Artical [sic] of the Statutes of the IAU, it is the Chinese Society of Astronomy only which has the right to represent China at the IAU.

[This refers to the 'Statement' of November 1974 by CAST mentioned above; Appendix 2.4.]"

Then came the General Assembly (its XVIth edition), hosted by the Grenoble University. A much-awaited event was the opening address by Goldberg, on August 24. After having exchanged thanks and greetings with the French officials, in fluent French, Goldberg addressed the Assembly to update the participants on the progress about his commitment at the Sydney General Assembly, three years before, to bring back China into the Union: "[for the Executive Committee, no task] *is more urgent than finding some way to effect the return of our colleagues from China to the IAU.*"

Although the attentive reader is by now well aware of the efforts made by Goldberg and his colleagues to that end, it is good to read some excerpts reflecting, in his own words and in public, over the situation as it was by the end of 1975, 15 years after China had withdrawn from the IAU:

> I should like now to speak about the problem of Chinese membership in the IAU, which is a matter of the utmost concern to all of us. (...) Unfortunately, I cannot report that we have been successful in this endeavour, although we, together with ICSU, have made earnest efforts to dissolve the deadlock in which we find ourselves. (...) Considering that the membership of the IAU has tripled since China's withdrawal in 1960,[38] it is obvious that most of those present here will not be familiar with the issues relating to the Chinese problem and the attention being given by the Executive Committee to its solution, and therefore a few words of explanation are in order. (...)
> May I point out at the outset that the IAU has never regarded Taiwan as in any sense substituting for the People's Democratic Republic of China. (...)

---

[38] More precisely, from 1289 individual members at the Berkeley General Assembly (1961) to 3805 at the Grenoble General Assembly. The respective numbers of participants were almost in proportion: 765 and 2184 (Andersen et al., 2019). A graph showing the exponential growth of the IAU over a century since its creation in 1919 can be found in Montmerle (2019: fig. 5, p. 298).

Why did China leave the I A U? Because, if I may quote from the joint statement issued at the conclusion of President Nixon's visit to Peking in 1972, the Chinese government 'firmly opposes any activities which aim at the creation of 'one China, one Taiwan', 'one China, two Governments', 'two Chinas', and 'independent Taiwan' or advocate that 'the status of Taiwan remains to be determined''. They insist that Taiwan is a province of China and therefore that the Academia Sinica of Peking must be recognized as the sole adhering organization representing all of China. Furthermore, they would require the I A U to deny admission to representatives from Taiwan to all I A U sponsored conferences. For the I A U to accept these conditions would not only violate both its principles and statutes, but would establish a precedent that could lead in the future to the politically motivated expulsion of other members of the Union. (. . .)

Some of our colleagues are growing restive at the continued absence of Chinese astronomers from I A U meetings and would like quick action to speed their return. (. . .)

I would like to reiterate that, at present, the re-admission of China to the I A U would require a change in the statutes of the Union. (. . .) If the statutes were to be changed to allow cancellations of membership, the criteria would have to be clear and unambiguous. Two proposals have been made: (1) to restrict membership to countries that belong to the U. N. or UNESCO or (2) to require a certain minimum level of astronomical activity as a condition for membership. In my judgment, both of these proposals are in conflict with the aims of the Union and are therefore unacceptable, the first for reasons to which I have already alluded and the second because it would force more than one country out of the Union and would stifle the progress of astronomy in developing countries. (. . .)

Despite these difficulties, I believe that the present impasse is not insoluble and that we must continue to strive for an arrangement that does not require China to modify its political position towards Taiwan and yet at the same time makes it possible for astronomers on Taiwan to participate in the work of the I A U. (. . .)

One has to recall that Goldberg had organized the Berkeley General Assembly while being one of the earliest IAU officials to be aware of the application of Taiwan at the time of the Moscow General Assembly in 1958, then had been an IAU Vice-President for 6 years (1961–1967), before becoming President in 1973. So, in all 12 years of constant personal implication within the IAU Executive (not to mention his term as Advisor that would follow, from 1976 to 1979), spread over nearly 20 years. For Goldberg, addressing the General Assembly was a form of personal legacy to the fast-growing astronomical community, including astronomers in Taiwan, and of course, albeit indirectly, astronomers in China.

Not all participants, however, were convinced. In an article published in the Grenoble General Assembly daily Newspaper *"La Gazette d'Uranie"* (Fig. 5.11) on the last day of the conference (Tuesday, August 31), in spite after having had a private discussion with Goldberg, a group of astronomers (J.M. Greenberg and G.K. Miley[39] from Leiden, the Netherlands, R.S. Booth from Sweden, W.N. Christiansen[40] from Australia, R. Fanti from Italy, Tao Kiang from Ireland, and M. Rowan-Robinson from UK), most of them radioastronomers, attacked the IAU:

---

[39] Miley (2022).

[40] See above, and Wang Shouguan (2022).

Fig. 5.11 The Grenoble General Assembly daily newspaper, issue of August 31, 1976, and the article criticizing the IAU for having admitted Taiwan in 1959: "China and the IAU" (© IAU)

We believe that the decision by the IAU in 1958[41] to admit 'The Republic of China' of Taiwan to membership was a mistake. It is now generally admitted that the application of Taiwan for membership was politically motivated. It was part of a plan by the US State Department under Dulles to gain Taiwan's admission to as many international organizations as possible, thereby isolating the Peoples Republic of China internationally. By then admitting Taiwan (where there was a negligible amount of astronomical activity) to membership, we believe that the IAU unwittingly allowed itself to be used for political ends. (...) As a first step towards improving relations with China it is surely important to admit that the 1958 decision of the IAU was a mistake. (...) Further, we feel that the IAU should officially express its intention of working to insure Chinese representation in the International Cultural and Scientific Union (ICSU).[42] Those who consider our action as 'bringing politics into the IAU' should remember that by its very structure (adherence of countries), politics is an unavoidable part of our organization. (...) We urge all of you who do not wish to see this embarrassing unjust and unrealistic situation continue, to solicit support for the motion below. (...)[43]

---

[41] Correction: It was in 1959 (Executive Committee meeting at Herstmonceux, see Chap. 3).

[42] Correction: ICSU stands for "International Council of Scientific Unions."

[43] The motion and the article can be downloaded in full at: https://www.iau.org/static/publications/ ga_newspapers/19760825.pdf

Two days later, on September 2, the incoming Executive Committee held its first meeting (#42) in the Physics Building of the Grenoble University campus, and the article was put on the agenda, with comments from Goldberg and Contopoulos. Two modest decisions were taken: *"It was agreed that the President should, jointly with Ch. Fehrenbach, meet the Chinese cultural attaché in Paris,"* and *"As regards Taiwan, it was approved that the Union continue referring to it as 'Taiwan' leaving out the words 'Republic of China'."*

The issue would not be discussed at this meeting of the Executive Committee, but was followed up by the new IAU President, Adriaan Blaauw, the one who would "close the China–Taiwan case." Interestingly, the "case" had been open by another Dutch President, Jan Oort, 20 years before... and both belonged to the Leiden Observatory.

Chapter 6
The Question of Taiwan: Pressure from UNESCO

6.1 The UNESCO Resolution of 1973

In parallel with the discussions between ICSU officials, IAU astronomers, and Chinese astronomers, a new player came on the scene: UNESCO. And with a clear goal for its Director General, R. Maheu: in conformity with its status as a UN organization ("UN" and "O"), *to "urge"* (and not *to convince*) all non-governmental organizations with which it had relations to *expel Taiwan* as the UN had done in 1971.

Indeed, on October 31, 1973, and labeled "URGENT," the ICSU Secretary General F. Stafleu (having been notified himself unofficially, probably by Prof. Chang Wai, since they were having discussions in China as part of the ICSU Mission),[1] sent a circular to the "Secretaries of International Scientific Unions, of Scientific and Special Committees, Inter-Union Commissions and Permanent Services," forwarding them a Resolution adopted by the UNESCO Executive Board in a recent meeting (reproduced here in full, *verbatim*):

---

[1] Section 5.4.

During the recent meeting of the Unesco Executive Board the following resolution was adopted:

1.   Recalling resolution 2758 (XXVI) adopted by the United Nations General Assembly on 25 october 1971, to 'restore all its rights to the People's Republic of China in the United Nations and to expel forthwith the representatives of Chiang Kai-shek from the place they unlawfully occupy at the United Nations and in all organizations related to it',
2.   Remembering at the same time that the Executive Board at its 88ᵗʰ session (88 EX/Decisions 9) decided that the Government of the People's Republic of China is the only legitimate representative of China in Unesco,
3.   Observing with satisfaction that the Director-General had already informed all international non-governmental organizations maintaining relations with Unesco of the position of the Government of the People's Republic of China on this matter,
4.   Noting with close attention that bodies or elements linked with Chiang Kai-shek and usurping the name of China continue to conduct illegal activities within certain international non-governmental organizations maintaining relations with Unesco,
5.   Urges international non-governmental organizations which maintain relations with Unesco and in which bodies or elements linked with Chiang Kai-shek participate, having illegally usurped the name of China to take measures to exclude them immediately and to break off all relations with them;
6.   Invites the Director General:
(1)   to communicate this resolution to all international non-governmental organizations maintaining relations with UNESCO;
(2)   to request the international non-governmental organizations concerned to inform him of the action they have taken to comply with the resolution;
(3)   to submit a report on this subject to the Executive Board at its Spring Session in 1974.

No mention was made however of possible sanctions in case of non-compliance with UNESCO's "orders," but there was an obvious threat in the background, that of cutting funds to restive organizations. (At that time, UNESCO was still funding about 40% of the ICSU budget, see Fig. 4.1 above.) The seriousness of the Taiwan case grew one step further: UNESCO was now pressuring "international non-governmental organizations" (including its long-time partner ICSU itself), in particular disregarding their own rules (their Statutes).[2] In other words, the Taiwan-China case was pushing international law to its limits, initiating an unpleasant standoff between UNESCO, on the one hand, and ICSU and its Unions on the other, bodies which in principle collaborated and shared the same values, as warmly recalled by F. Stafleu to R. Maheu in his letter of February 2, 1973 (Sect. 5.3).

Indeed, to resist this pressure was exactly the opinion of Goldberg in a letter (November 26) to Contopoulos, the General Secretary:

---

[2] It is intriguing that during the same period the UN started other initiatives to rein in, or otherwise compete with, some Unions. This was the case of a *"UN Group of Experts on Extraterrestrial Names,"* created in 1971 and openly challenging the IAU authority on celestial nomenclature. This Group was however not part of UNESCO, but of the UN's Department of Economic and Social Affairs. See Montmerle (2022b).

To exclude Taiwan 'immediately' as UNESCO urges would be in clear violation of our statutes. It remains to be seen whether UNESCO intends to punish unions that do not accede to their wishes by cutting off financial support. The action of UNESCO is on the face of its political interference with the International Scientific Unions and my initial instinct is to oppose their request.

At the same time, Goldberg suggested, and Contopoulos agreed in his answer, not to circulate the UNESCO resolution to the Executive Committee until more was learned about the results of the ongoing ICSU Mission to China.

In parallel (on November 30), Goldberg wrote to three Chinese-American colleagues (C.N. Yang, from Stony Brook, already mentioned, C.C. Lin, from MIT, and C.Y. Fan, from the University of Arizona) to ask for advice. In his answer (December 7), C.C. Lin, obviously unaware of the UNESCO resolution, makes however an interesting connection:

I was in Peking six weeks ago. (...) My own impression is that [the Chinese] feel that things would eventually be straightened out via UNESCO. I presume that the idea is as follows. In the United Nations the delegation from Peking was recognized as representing China. It is therefore not a question of expelling Taiwan from the IAU, but a question of recognizing which delegation should represent the Chinese membership in the IAU. Among the Chinese people, as Kissinger again officially noted during his recent visit to Peking [in 1972], there is agreement that there is only one China, of which Taiwan is a part. If this position were taken by the IAU, the logical solution to the problem of Chinese membership would be quite simple and clear.

But Contopoulos, in a letter to Goldberg (December 14), disagreed:

Personally, I feel that we cannot accept the Chinese claim that 'China is responsible for its (Taiwan's) representation'. This would lead to endless claims from countries that do not recognize other countries, or from people that oppose strongly a given government. If, e.g., South Africa[3] is expelled from UNO does it mean that it should be expelled from ICSU and IAU also?

To end the year 1973, on December 26, Goldberg, following some suggestions, provided Contopoulos with a fresh idea: to find a convincing, "right" wording for Taiwan in the IAU.

We cannot accept the UNESCO request that Taiwan be expelled from the IAU since that would be in violation of our statutes." (...) Although we very much appreciate the subvention from UNESCO, we must insist on complete independence from political organizations of any kind. (...) Despite my attitude toward UNESCO, namely, that the IAU should resist political pressure, it is obvious that we cannot escape some consideration of political realities. (...) When we remove the phrase (Republic of China) from the listing of Taiwan's membership in ICSU and the IAU, we are in effect making a political judgment. Therefore I do not see why we should not go further in this regard and *refer to Taiwan as a province of China,* the important point being that in this way we still make it possible for astronomers in Taiwan to be represented in the IAU.

---

[3]Then under the Apartheid regime; see above, Footnote 20 in Chap. 5.

Note that this was actually a step back from his previous suggestion ("part of China," instead of "province of China"), perhaps in a conciliatory hope to be more agreeable to the Chinese.

In any case, there was apparently no answer from Coulomb to UNESCO until the end of his term, which was not completely unexpected since he would step down as ICSU President only a few months later, at the Istanbul General Assembly (September 1974). But the story was far from over.

6.2 UNESCO, Act II (1975)

In the absence of an answer from ICSU, the following months did not bring particularly good news from UNESCO toward a solution of the China conundrum. On the contrary UNESCO came back rather aggressively with a reminder to those bad pupils among the Unions who did not comply with UNESCO's firm request for information about their actions to expel Taiwan from their members, in the form of a new Resolution.

This Resolution, dated November 23, 1974 (#6.51 of the 47th UN plenary meeting, to be precise), was essentially identical to that of October 31, 1973 (above), but updated with a new deadline from UNESCO's Director General, now Amadou-Matar M'Bow, to produce a report on the situation: autumn 1975.

So, on January 9, 1975, A-M. M'Bow sent a letter to "International non-governmental organizations maintaining official relations with UNESCO," as before, urging them to send him, before June 2, "as detailed information as possible on the points raised in the resolution." The fact that M'Bow did not mention a previous reply by ICSU shows that, indeed, there had been none from Coulomb, but his successor, Harrison Brown (see Footnote 19 in Chap. 5) now took measures to answer him, starting with a wide consultation of the Unions.

On March 17, Brown issued a two-page circular to the Presidents of all ICSU Unions (here to Goldberg for the IAU), developing his views about the role of ICSU as a basis for receiving feedback: In short, whether or not the Unions would support expelling Taiwan from ICSU. Knowing the IAU's position, it is interesting to give large excerpts from this document:

> In light of the increasing political pressures being felt by international non-governmental organizations, I thought it would be helpful to share some thoughts with you and in turn solicit your views.
>
> As non-governmental organizations, ICSU and the Unions are concerned with advancing the various scientific disciplines for the benefit of all peoples through international cooperation. Over the years they have established a far-reaching pattern of cooperation and trust transcending the political barriers that divide the world.
>
> (...)
>
> Achieving full participation and membership of colleagues in the People's Republic of China in ICSU and the Unions is a goal I am sure we all share. (...) But I also feel strongly that we would do ICSU and the Unions irreparable harm in the long run if we were to violate our principle of universality and our non-political character by casting out one of our members as a political precondition to another's applying for membership.

I believe it is proper, however, for ICSU or a Union to insist that a National Member not pretend to represent scientists in geographical areas over which it has no effective control. As you may know, when I visited Taiwan last year [1974] for discussions of this question, it was formally stated by the President of their Academy that '...*insofar as ICSU and its Unions were concerned, the scientific community and the scientific endeavor represented by the Academia Sinica are only those in the geographic area coextensive with that under the effective administration of the Republic of China*'.

(...)

We must exert a greater effort toward convincing governmental and inter-governmental organizations which attempt to use ICSU for political ends of the soundness of those principles and our evenhandedness in our applications of them.

As an aside, I have outlined above in italics the key sentences (in slightly different words) of the IAU Statutes (Article 3b) that were used by its Executive Committee to admit Taiwan in 1959,[4] but here expressed by Taiwan itself to ICSU—an interesting turn of events indeed!

Having received early answers, Brown did not want to wait any longer to respond to M'Bow, repeating in essence the arguments of his circular (May 14), but also explaining ISCU's efforts of concertation with the Chinese Academy of Sciences:

... [the Unions] who have answered uniformly agree that we could not take the position of casting out one of our members as a political precondition to another's applying for membership. (...) I would hope neither ICSU nor UNESCO would jeopardize their mutually beneficial relationship by trying to impose their interpretation upon the other.

However, as it turned out, later answers were not all unanimous: Brown reported to the Union Presidents (July 10):

... I was pleased to learn of the earnest efforts that several of the Unions have made to establish contact with scientific colleagues and institutions in the People's Republic of China.

In general, the replies indicated agreement with the views presented in the letter [March 17]. There was also dissent. (...) On the other hand, one of the purposes of that letter was to stimulate discussion and to elicit opinions. Therefore, I would be very grateful to hear in the weeks ahead from those until now silent.

The "dissent" was expressed in a letter from the URSI Vice-President H.G. Booker. This could have been alarming, since URSI (French acronym, still in use, for *Union Radio-Scientifique Internationale*, translated as the *International Union for Radio Science*) was one of the founding "historical" Unions of the 1920s,[5] thus one of the most important Unions within ICSU. Although the Booker letter could not be found in the URSI archives, a purely administrative exchange between C.M. Minnis, URSI Secretary General, and H.C. Fang, President of the URSI Committee of the Republic of China (Taiwan), resp. on June 5, 1975, and March 1, 1976, shows that his country normally paid its annual contribution for 1976 (of US$300, i.e.,

---

[4] Section 3.6.

[5] Fauque and Fox (2022).

US$1320 today).[6] So, whatever the internal discussions within the URSI Executive (of which their archives have no record), the conclusion was that URSI would eventually follow the ICSU policy, and that they would not expel Taiwan.

6.3   The ICSU Dissidents

Dissenting actions did indeed take place, however. At its 1975 IAU Executive Committee meeting (#40), held on September 9–11 in Lagonissi (Greece) at the invitation of Contopoulos, the latest news from China were discussed. Item 7 of the minutes reports:

> ... It was noted that IUGS resolved that the admission of Taiwan was made in breach of their statutes. The IAU should not follow the action taken by UNO. In fact, the IAU had accepted the membership of the People's Republic of China long before this country was accepted in the UNO. The IAU will go as far as possible to meet the objections of China, but without changing its principle that astronomers from all parts of the world should be allowed to participate in it.

The IUGS (*International Union of Geological Sciences*) was founded in 1961 and had adhered to ICSU at the same time. Today, it represents about 1 million geologists and Earth scientists and has 120 National members, and its Secretariat is located in Beijing (Peking). Pending further investigations, it is likely that it was one of those Unions which Taiwan had joined in the same "wave" as the IAU and shortly thereafter. Its ties with UNESCO were strong, with the International Geological Programme started in 1972 (Greenaway, p.186). This may be the reason why it decided to switch from Taiwan to China in 1976.

Another dissident was the IUGG (*International Union of Geodesy and Geophysics*), one of the "historical" Unions. The way it opposed ICSU drew the ire of Harrison Brown, who wrote to Sir John Kendrew (the new ICSU Secretary General), on July 13, 1977:

> I have received your telex concerning the Draft Report of the last ICSU Executive Board meeting and the wording of Item 7 on our discussion of the IUGG motion on China. I continue most strongly to object to the use of the word consensus which implies unanimity or general agreement. (. . .) The Executive Board minute as you are prepared to distribute it, represents a major change of ICSU policy. (. . .) In my opinion, the Executive Board has transcended its authority in this case. This action, as now worded, is in direct contradiction with past ICSU resolutions, most notably on political nondiscrimination of 1958, and on universality of 1976. I want to emphasize again that such action at the very minimum requites a formal vote.

Another letter in the same vein was sent on July 29 to the Members of the ICSU General Committee, asking *"that the entire question be placed on the agenda for the General Committee* [September] *meeting in Budapest. It is to be most sincerely*

---

[6]I am indebted to Pierre Baüer, former URSI President (1993–1996), and to Inge Heleu, Administrative Secretary of URSI, for conducting this research for me in the URSI archives.

regretted that the Durham meeting [on August 6] *of the IUGG will place before this major policy change can be considered by the all members of the ICSU family and with the impression that the proposed IUGG action has the endorsement of ICSU. I suggest it is still to be determined."*

Alas for Brown, the IUGG proceeded defiantly: On August 11, P. Melchior, IGGU General Secretary wrote (in French): *"I inform you that the People's Republic of China has become a member of IUGG on August 6, 1977. The representation of the Republic of China (Taiwan) to the Council of the Union has been cancelled. This decision was taken during the Council by 47 votes in favour, and 7 against."* The reason given in the motion was only to follow the UN decision, in other words to comply with UNESCO's request.

In reaction to this *coup de force,* the IAU Executive Committee, at its following meeting (#43, held in Geneva Observatory, August 29–September 2, 1977, under the presidency of A. Blaauw and at the invitation of General Secretary E. Müller), reiterated: *"The Executive Committee agreed that the IAU, if confronted with the problem of expelling Taiwan from among its ranks would see no reason to deviate from its policy as stated by Prof. Goldberg's address to the XVIth General Assembly* [Grenoble, 1976, see Sect. 5.7]. *The General Secretary was authorized to express this stand of the IAU to the General Committee of ICSU, if necessary."*

In 1978, in a letter from Chou Pei-yuan, Vice-President of the Academia Sinica in Peking (who, as we have seen, already met ICSU's President J. Coulomb in 1973) to Kendrew (September 10), there is the mention of another Union, IUCr (*International Union of Crystallography*) having admitted China. As it turns out, the IUCr is one of the few Unions that post a detailed account of their history, written by H. Kamminga in 1989.[7] This Union became a member of ICSU in 1947. Apparently, Taiwan was not a member, and this represented another scenario. Quoting from the report:

> The political realities of the world have also had their impact with respect to adherence to the IUCr. For example, for many years the People's Republic of China wished to adhere, but only on condition that the Union would never admit Taiwan. The IUCr resisted such demands and when the People's Republic eventually joined in 1978, it was admitted to the IUCr on the stipulation that this act would not prejudice the future admission of any other country.

So in this case, even though Chou's letter lists IUCr alongside the bona fide ICSU "dissidents," in reality it was not one…

All in all, in view of the relatively large number of Unions within ICSU (about 20 at that time), and although they admittedly represented a large number of scientists, the number of dissidents was in fact small, albeit not to be taken lightly (IUGG, IUGS), and they took their decision very late in the process, several years after UNESCO had "urged" ICSU and its Unions to expel Taiwan. The events of the late 1970s to early 1980s would eventually solve "naturally" the remaining China–Taiwan membership issues.

---

[7] https://www.iucr.org/iucr/history/early-history

UNESCO, on the other hand, seemed to have understood the message, and eventually did not break off from restive Unions. At the time of its quarrel, as we have mentioned ICSU still received 40% of its budget from UNESCO; as for the IAU, the minutes of the Executive Committee meeting (#43) at Geneva Observatory in September 1977 indicate (item 17):

> The General Secretary said the UNESCO had discontinued to award grants to the Unions and prefers now to subsidize specific projects in the form of contracts. The Union [the IAU] has therefore applied within the UNESCO programme for 1979-1980 and the Medium-term Plan 1979-1984 for three contacts (. . .).[8]

---

[8] The total amount requested was $12,300 per year for three Commissions (5, 44 and 46), or $54,000 in 2020 US dollars. Commission 5 was "Documentation & Astronomical Data," one of the IAU's first Commissions (created in 1922); Commission 44 was "Space & High-Energy Astrophysics," created in 1958; and Commission 46 was the well-known "Astronomy Education," created in 1964 and of obvious interest for UNESCO.

Chapter 7
The Uncertain Road to Reconciliation

7.1 "Melting the Iceberg"

At the end of the Grenoble General Assembly, a new Executive Committee opened the 1976–1979 triennium by a meeting on September 2, 1976 (#42) in the Physics Building on the campus of the Grenoble University. The new President was Adriaan Blaauw, from Leiden, the Netherlands (like his compatriot Jan Oort). Born in 1914 and educated in Leiden and Groningen like Oort, he had just finished a five-year term as ESO Director General, and he had done seminal work on high-mass star formation, star clusters, and stellar associations in the Milky Way.[1]

The new General Secretary was Edith A. Müller, from the Geneva Observatory, a specialist in solar spectroscopy.[2] The other members were the second-term Vice-Presidents John Bolton (Australia), Charles Fehrenbach (France), and Wilhelmina Ivanovksa (Poland), and the new Vice-Presidents were David Heeschen (USA), Evgeni Kharadze (Georgia, USSR), and Sydney van den Bergh (Canada).[3] Of these, the most active members on the China problem, in addition to L. Goldberg and to some extent G. Contopoulos, now advisors, would definitely be A. Blaauw (Fig. 7.1) and E. Müller (Fig. 7.2), as well as Ch. Fehrenbach.

Toward the end of Chap. 5 above, we quoted excerpts from L. Goldberg's opening address, on August 24, 1976, in Grenoble. While these excerpts centered on the "formal" relations between China, Taiwan and the IAU, the address actually

---

[1] See his lifetime achievements on the occasion of his 90th birthday (April 12, 2004) on the website of the University of Groningen (in Dutch): http://www.rug.nl/sterrenkunde/onderzoek/Blaauw/blaauwLeven

[2] Edith Alice Müller (1918–1995). See her tribute book (Appenzeller et al., 1998).

[3] Dates for the first-term Vice-Presidents: D. Heeschen (1926–2012), E. Kharadze (1907–2001), and S. van den Bergh (1929–). The second-term Vice-Presidents were part of the previous Executive Committee (Footnote 27 in Chap. 5).

© The Author(s), under exclusive license to Springer Nature Switzerland AG 2022
T. Montmerle, Y. Zhou, *China and the International Astronomical Union*, Historical & Cultural Astronomy, https://doi.org/10.1007/978-3-031-01787-2_7

Fig. 7.1 *Left:* Adriaan
Blaauw (1914–2010), at the
time of its General
Directorship of ESO
(1970–1975). (© ESO)

Fig. 7.2 *Right:* Edith
Müller, with students at her
Geneva Observatory office.
(Courtesy G. Meynet,
Geneva Observatory)

ended, on a more engaging tone, with an encouragement to all astronomers to act "informally," by furthering collaborations with their colleagues from the Mainland:

> Irrespective of whether China rejoins the IAU in the immediate future, we astronomers should try to accelerate the frequency of contacts with our Chinese colleagues and to develop cooperative projects on an individual or country-to-country basis. There are now many signs that the Chinese have resumed the expansion and development of their activity in astronomy and that cooperation and exchanges with astronomers of other countries would be welcomed by them. (...) I would also urge astronomers, as the General Secretary urged all National Committees one year ago, to use every opportunity to invite Chinese astronomers to conferences in their own countries and to let them know how eager we are to communicate and work with them.

Now simply an IAU advisor, Goldberg, still Director of NOAO in Tucson, Arizona, quickly applied these recommendations to himself. At his invitation, Wang Shouguan (whom we already met in his connection with the Australian radioastronomer W.N. Christiansen),[4] Director of the Radioastronomy Department

---

[4] Section 5.1 and also Wang Shouguan (2022).

of the Peking Observatory, and a group of about ten Chinese astronomers, visited several observatories and institutes in the USA (letter to E. Müller, November 3, 1976): NRAO and the VLA,[5] but also Harvard and MIT,[6] of course Tucson and Kitt Peak, and others. A very complete tour indeed!

In return, Goldberg and a delegation of nine American scientists, including eight top astronomers and a Professor of Chinese Culture and of the History of Science from Philadelphia, were invited on a month-long trip during the fall of 1977 to visit major Chinese cities and observatories (Peking, Shanghai, Nanking, Kunming, Kweilin and Canton).[7] Upon returning, the delegation published a 110-page report entitled *"Astronomy in China,"* under the auspices of the NAS Committee on Scholarly Communication with the People's Republic of China (Goldberg and Edwards, 1979; Fig. 7.3). This was the first time such a report was made available to the public as the fruit of an official cooperation between American and Chinese institutes in astronomy.

Other exchanges were organized, for instance between France and China: The minutes of the Executive Committee meeting #43 (Sauverny, Switzerland, August 29–September 2, 1977) report an invitation of Ch. Fehrenbach to China,[8] (Figs. 7.4 and Fig. 7.5) and a letter from him (October 16, 1978) to E. Müller mentions a return visit by a delegation of five Chinese astronomers to the Observatoire de Haute-Provence in southern France. Here, having questioned his guests about the return of China to the IAU, Fehrenbach got no reply and concluded (in French as always): *"Toute cette affaire paraît évoluer bien difficilement... Le plus simple est de ne rien faire et d'attendre."*[9]

Another illustration of fruitful contacts between French and Chinese astronomers was given when Ye Shuhua, then Vice-President of the Chinese Astronomical Society and future member of the Chinese delegation at the Montreal General Assembly the following year,[10] visited France, also in 1978, as head of a delegation of the Chinese Academy of Sciences (CAS). A specialist of astrometry, Ye Shuhua met French experts in Paris and Nice; in return, at her initiative, the CAS extended an invitation to a French delegation, which visited China in 1979. This visit initiated a long-time collaboration: The live story, by Pierre Léna from Paris Observatory, who was part of this delegation, is told in Appendix C.

---

[5] NRAO: National Radio Astronomy Observatories (Charlottesville, Virginia); VLA: Very Large Array radiotelescope (near Socorro, New Mexico).

[6] Harvard College Observatory, and the closeby Massachusetts Institute of Technology, near Boston.

[7] Now Guilin (桂林) and Guangzhou (广州), respectively.

[8] Fehrenbach gives an account of this visit in his autobiography (Fehrenbach, 1990, pp. 507–517). The story of the Xinglong Observatory is told in Appendix C of Montmerle et al. (2022).

[9] "This matter seems to evolve with quite some difficulty... The simplest is to do nothing but wait."

[10] See below, Sect. 7.3.

Fig. 7.3 Purple Mountain Observatory (PMO) visit, part of the tour in China organized by L. Goldberg, October 16, 1977. The American delegation and party with hosts: *First row:* Wu Ling-an, interpreter, Institute of Physics, Chinese Academy of Sciences; Teng Ting-yu, staff, Scientific and Technical Association of the PRC (STAPRC); Kung Shu-mo (PMO); Hsing Te-lin, PMO; Margaret Burbidge (University of California, La Jolla); Martin Schwartzschild (with glasses; Princeton University Observatory); Tseng Wu-shu, PMO Secretariat; Victor M. Blanco (Cerro Tololo Inter-American Observatory, Chile). *Second row:* Li Ming-te, Staff, STAPRC; Ts'ui Lien-shu, Nanking University; Chao Wen-piao, PMO Administrator; George Herbig (with glasses; University of California, Santa Cruz); Leo Goldberg (Kitt Peak National Observatory); Harlan J. Smith (McDonald Observatory); Allan Sandage (Hale Observatories); David S. Heeschen (National Radio Astronomy Observatory). *Back row:* Chang Yu-che (Zhang Yuzhe), PMO Director (with cap); Nathan Sivin (Professor of Chinese Culture and of the History of Science, University of Pennsylvania); Richard Bock, US Liaison Office in Peking; Charles H. Townes (University of California, Berkeley) (© National Academy of Sciences)

In another context, W.N. Christiansen and Wang Shouguan resumed their collaboration dating back from the early 60s, etc. No doubt other exchanges took place, but since, by definition, they were not acceptable by the Chinese if carrying an IAU stamp, those not involving Executive Committee members were in general not recorded in the archives.

Following Goldberg, there were also attempts at inviting Chinese astronomers to IAU "Regional Meetings" (which are more focused on regional exchanges than thematic Symposia, or Colloquia—now discontinued): For instance, an invitation was sent to Y.C. Chang to attend an Asia–South Pacific Regional meeting to be held in New Zealand in April 1977, but Chang politely turned it down because it was an official IAU meeting. Thus, no progress in this case.

Fig. 7.4 Visit of
Ch. Fehrenbach to the
Xinglong Observatory (July
1977). Meeting with his
long-time friend Cheng
Maolan (Director of the
Observatory, with
sunglasses; see Sect. 5.1)

Fig. 7.5 Fehrenbach lecturing students (Source: Video program about Cheng Maolan shown on Hebei TV in China: https://v.qq.com/x/search/?q=%E7%A8%8B%E8%8C%82%E5%85%B0& stag=0&smartbox_ab=)

In February 1978, a Chinese delegation attended a CCIR[11] meeting in Geneva. It was the first time ever that such a delegation, which included two astronomers, attended one of these meetings. They expressed the will to take this opportunity to meet with E. Müller, the IAU General Secretary, since she was based in Geneva. She reported that the Chinese were open to rejoining somehow the IAU, but on the condition that Taiwan would not be a member anymore (as could be expected from a UN agency anyway), citing the example of the IUGG and IUGS. While Müller reiterated the IAU position, as expressed by Goldberg at Grenoble, the meeting was courteous, and the head of the delegation, Mr. Yang Chao-chin, a Peking government official, concluded, coincidentally echoing Fehrenbach: *"We can wait."*

Yet the iceberg was beginning to melt slowly. Following the Geneva contacts, as reported by Fu&Ye:[12]

> Several months later, with the invitation from the Dutch Academy of Sciences, the Chinese Academy of Sciences delegation visited the Netherlands. During this visit, the Chinese officials had an extensive opportunity to further discuss this matter with the IAU officials. That was because the President of the IAU at the time Adriaan Blaauw was from the Netherlands. He was a professor at the Leiden University as well. While the Chinese were visiting the Leiden Observatory on July 13th, Blaauw gave a speech which included contents about the European Southern Observatory, astronomy, astrophysics and the IAU's international relations. His message behind this speech was quite obvious in the Chinese delegation's eyes. During the various occasions following suit, including a banquet held at the Chinese Embassy in the Netherlands, Blaauw exchanged his point of views with Chinese officials, including Zhu Yongxing (1930–), who was the administrative vice bureau chief of the Chinese Academy of Sciences Foreign Affairs Bureau. Blaauw was given the impression that the Chinese officials asked the IAU to wait for the right opportunity and the de-escalation of both sides opposition of views.

The situation was also evolving positively on totally different horizons. Starting on August 23, 1978, a High-Energy Physics conference took place in Tokyo. A newspaper clip found in the IAU archives (not referenced, but authored by a *Washington Post* journalist, Jay Mathews, a China expert)[13] ran the headlines: *"Ending 30 Years of Boycott: China, Taiwan Delegates Attend Same Conference."* In the article, a diplomat is quoted as saying: *"The Tokyo meeting is another sign of China's new flexibility on foreign policy,"* noting *"Peking's new treaty with Japan and Chairman Hua Kuo-feng's[14] trip to Eastern Europe."* The article also gives important clues to the new Chinese foreign policy *vis-à-vis* Taiwan:

---

[11] *Comité consultatif international des radiocommunications* (International Consultative Committee for Radiocommunications): Based in Geneva, it was a United Nations agency since 1947. It was replaced in 1993 by ITU-R (*International Telecommunications Union-Regulations*). It is important for radioastronomy.

[12] Actually, this excerpt from Fu&Ye follows very closely the account given by Blaauw in his book (p. 199), but adds some new information from the Chinese side. See Appendix D for the full text.

[13] https://en.wikipedia.org/wiki/Jay_Mathews

[14] Hua Kuo-feng (now Hua Guofeng, 华国锋) was the appointed successor of Mao Zedong, who had died on September 9, 1976; see Sect. 7.2.

Attending the conference may represent even more of a departure for Taiwan than for China, for Peking in recent months has given several signs of a new approach to dealing with its Taiwanese adversaries. Rep. Lester Wolff, D-N.Y., said last month after visiting Peking that the Chinese had expressed a new willingness to negotiate directly with the Taiwanese. The offer seemed mostly an effort to win sympathy for Peking in the United States, since Taiwan is unlikely to consent to such talks.

A paragraph of the article further down is more directly related to the new situation in science: *"China's presence at the Tokyo conference indicates deep interest in reviving scientific research and establishing contact with foreign scientists after several years in which political struggles left little time for science."*

Just a month later, ICSU held its 17th General Assembly in Athens. In a Statement given on September 27, Dr. Ta-You Wu, Chief delegate of Academia Sinica in Taipei, declared: " *(...) With the actions of IUGS and IUGG* [to expel Taiwan], *what is involved is not the accessibility of scientific meetings to an individual, but rather the inherent right of an independent scientific community to participate in union management. This right is protected by the Universality Principle* [of ICSU[15]]." (...)

We have repeatedly affirmed that we represent only the scientific community we actually and effectively represent, *and have now offered to add in parenthesis the word 'Taipei' or 'Taiwan' to the listing of all our adhering bodies.* This will remove any doubt whatsoever about our geographical representation on the one hand, and preserve our right to name our own institutions on the other. We believe our position is a reasonable one.

Note here the progress made by Taiwan: There is no request to include "Republic of China" in the listing, but simply a geographical reference. This would prove a key point (albeit not the only one) in the direct negotiations that were about to take place during the following year.

7.2   A Rapidly Evolving Geopolitical Context

It is perhaps appropriate now to make a brief detour toward the political developments in China and Taiwan at that time, because their chronology is crucial to understand the quick succession of events that took place in 1979, in the aftermath of the deaths of the two pillars of the People's Republic of China, Chairman Mao Zedong (September 9, 1976), and his Prime Minister Zhou Enlai (who had died a few months before, January 8, 1976), themselves following that of Chiang Kai-shek in Taipei (April 5, 1975).

On August 12, 1977, the XIth Congress of the Chinese Communist Party opened. The troubled succession of Mao was of course at the center of the agenda. After having been stripped from all his functions inside and outside the Party in 1968, rehabilitated in 1973 (Xth CCCP), then removed again in 1976 shortly before Mao's

---

[15] See Chap. 8 of Greenaway (1996): "The free conduct of science."

death, with his consent and supported by the "Gang of Four". Eventually, Deng Xiaoping was re-established in his capacity of Vice Chairman of the CCP in 1977. Hua Guofeng, appointed by Mao himself as his successor, was then elected to the combined roles of Chairman and Prime Minister, and Deng gave the closing address to the Congress, demonstrating his renewed influence. In December 1978, at the Third Plenum of the XIth Central Committee, Deng announced the launch of the *"Four Modernizations"* (四个现代化) program. This program had been initiated by Deng in the aftermath of the XIth Congress to completely overhaul China's economy, with radical reforms in "agriculture, industry, defense, science and technology."[16]

But such a bold modernization could not be achieved by China alone, and in stark contrast to the Mao era (like the disastrous "Great Leap Forward" of 1958), Deng decided to solicit international support, applying for financial and technical assistance from the UN Development Program in the fall of 1978. In other words, after the upheavals of the Cultural Revolution, China was economically and diplomatically opening up to the western world: It was in this context that the Tokyo physics conference discussed in the preceding section took place.

Another major international event occurred that would play a decisive role in solving the "China–Taiwan problem," in particular for scientific unions: On January 1, 1979, under the Carter administration, the United States established diplomatic relations with the PRC, and broke off its official relationship with Taiwan at the UN and in other organizations. Almost immediately afterward, on January 29, Deng started a tour of the United States, a week-long visit that concentrated on industry and technology. The embassies of both sides opened on March 1.

The situation also drastically evolved in Taiwan. Chiang Kai-shek had died four years earlier, so strictly speaking the "Chiang Kai-shek clique," so often mentioned previously in official Chinese (and UNESCO) documents did not exist anymore as such. However, the Chiang family was still firmly in power: His son Chiang Ching-kuo, born in 1910 to his first wife, Mao Fumei, and educated in the USSR, had been Taiwan's Premier since 1972. He became Chairman of the Guomindang[17] at the death of his father and was elected (by the National Assembly) President of the "Republic of China" on May 28, 1978 (and re-elected in 1981). Nevertheless, after January 1, 1979, the relations between the USA and Taiwan completely changed, being ruled simply by the unilateral "Taiwan Relations Act" approved by the US Congress. The last US troops left the island on May 3, and the US Navy Seventh Fleet stopped patrolling the Taiwan Strait.

Consequently, after its expulsion from the UN in 1971, Taiwan's position on the world scene as of 1979 became more difficult, encouraging a certain conciliatory profile that it would demonstrate in future negotiations, as well as showing a slow progress toward democracy.

---

[16] See, e.g., the "Perspective Monde" website (in French: Université de Sherbrooke, Québec, Canada), here https://perspective.usherbrooke.ca/bilan/servlet/BMEve/1150 and associated articles.

[17] "Guomindang", also referred to as "Kuomintang" (KMT), stands for "Chinese nationalist Party" (中國國民黨). Founded by Sun Yat-sen in 1912.

Concerning the China–IAU relations, Fu&Ye summarize the situation in this way:

> [After 1978], the situation was quickly evolving. The PRC's national policy in regards to its participation in international organizations changed ingeniously as well. Some policies were also adjusted in Beijing. Under the new policy guidelines, *the PRC's participation in international organizations no longer firmly demanded the eviction of the Taiwan delegation.* They only stressed the idea that Beijing would not accept 'two Chinas' or 'one China, one Taiwan'.

7.3   Speeding Up: The Montreal Opportunity

In spite of the fact that the events just described are mentioned only briefly in Fu&Ye and not at all in Blaauw's book, it is clear that they played a decisive role in the background, not only for the IAU, but also for ICSU and all the Unions having refused to expel Taiwan (see next Section below).

For the IAU, it turned into a race against the clock, skillfully managed by Blaauw as its President, because the next General Assembly was due to take place in Montreal only a few months later, on August 14–23, 1979. This would offer an ideal "window" to press both sides (China and Taiwan) simultaneously without delay to the table of negotiations, aiming at crafting a resolution for submission at the General Assembly, to the effect of quickly re-admitting China while retaining Taiwan.

However, Blaauw soon realized that the main hurdle, this time, would not be China, already open to a "cohabitation" with Taiwan in some adequate form for the political reasons mentioned above, in particular by the *Washington Post* journalist, but Taiwan itself, which, while having the institutional support from the IAU, was struggling to keep a responsible role on the international scene. Blaauw's key idea was to deal with China and Taiwan at the same time, and even to encourage them to arrive at a mutual agreement, on the basis of suggestions made by the IAU.

On April 9, 1979, after having solicited the advice of the Executive Committee, Blaauw sent a letter to Chien Shih-liang, the President of the Academy of Sciences in Taipei (a chemist), presenting three options aiming at allowing delegations of astronomers of both China and Taiwan to participate in the Montreal General Assembly. Pressing Chien, Blaauw wrote:

> For the IAU as an organization for which worldwide contacts and collaboration are exceptionally important, it is urgent that a solution be found, *even if only of a provisional nature,* to allow astronomers from both parts of China to soon become again properly involved in its work. This should preferably be the case <u>already at the forthcoming General Assembly</u>. I therefore wish to consult now with the Academy of Taiwan about alternative possibilities to achieve this, in the hope and expectation that *both the responsible authorities at Taipei and at Peking* will favorably consider ways to arrive at a reasonable and fair arrangement.

(The italics are mine; the underlined parts are in the original.)

Blaauw then elaborated on three "alternatives," which I summarize here as follows (see also his book, p. 199): (a) "The most desirable arrangement" would be a joint Chinese delegation established after mutual consultations of the Academies of Sciences in Peking and in Taipei; (b) the Academy of Taiwan acts as an observer only in all official matters which are statutory obligations of the adhering organizations (e.g., refraining from voting on the IAU budget and finances), while normally attending the scientific sessions, and symmetrically Chinese astronomers are admitted as observers; (c) consider astronomers from Taiwan *and* from Peking as individual Chinese IAU members, at least temporarily, in other words consider those who would attend the Montreal meeting as individual participants (i.e., not as "nationals"; essentially in the same way as physicists from Chinese and Taiwanese research *institutes*, not from "China," had participated in the 1978 physics conference in Tokyo).[18]

In parallel, the outside world was also on the move. In a newspaper clip sent by L. Goldberg to E. Müller, Blaauw and Wayman, and recorded in the archives (unreferenced American journal; place and date: "Montevideo, Uruguay, April 7" [UPI]); IAU Secretariat stamp in Geneva: arrival date 13 Avr. 1979), an article is entitled "*I.O.C. Votes China Compromise,*" starting with:

> The International Olympic Committee voted, 36-28, today to recognize the Chinese Olympic Committee in Peking while at the same time retaining Taiwan as a member. Lord Killanin, I.O.C. President, said he understood that Peking would accept a separate Olympic committee from Taipei 'as an interim measure'.

This document comes with a handwritten annotation by Goldberg (Fig. 7.6), obviously echoing the context of the Winter Olympics of 1960, while he himself was preparing the 1961 Berkeley General Assembly. . .[19]

But the story was not over: From Fu&Ye, we learn that the final agreement signed by the IOC in October 1979 was largely due to the C.S. Shen, the same Shen who would be instrumental in devising the IAU Montreal agreement itself (see below)!

As for Unions, on April 23, a Circular from E.C. Slater, the Treasurer of the International Union of Biochemistry (IUB)[20] to its Executive Committee contains a number of highly relevant remarks:

> I was informed that the Chinese Biochemical Society will be constituted, at a scientific meeting to be held in Hangchow[21] on May 22–30. Representatives from Taiwan have been invited to attend this meeting, and one of the ten positions of Vice-President and one of the 50-odd positions on the Executive Committee have been reserved for representatives from Taiwan. Immediately after its formation, the Chinese Biochemical Society will make an application to join IUB, attaching the condition that separate membership of Taiwan cease.

But Slater added: "*Our statutes do not permit the expulsion of an Adhering Body, except for a serious cause. Since the Academy of Science, Taiwan, represents a scientific community, as defined by the Statutes, no such serious cause exists at*

---

[18] Section 7.2.

[19] Section 3.3.

[20] E.C. Slater was from Amsterdam, which probably explains why this document came to the attention of A. Blaauw and ended up in the IAU archives.

[21] Now Hangzhou (杭州, formerly Hang Tcheou, Zhejiang province).

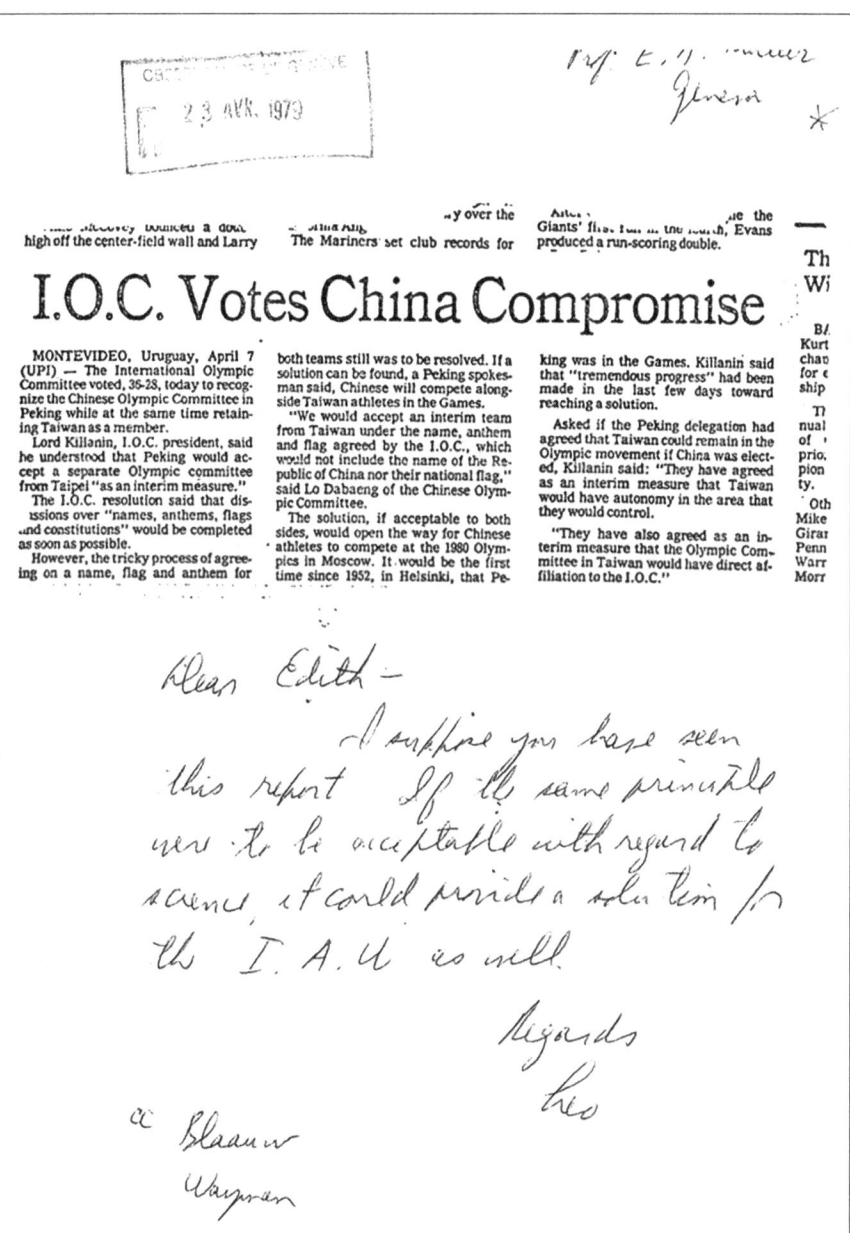

Fig. 7.6 Newspaper clip about China rejoining the International Olympics Committee, received in Geneva on April 23, 1979, with a handwritten comment by Leo Goldberg to Edith Müller, Blaauw and Wayman noting the similarity with the IAU (IAU archives)

present." So the problem, identical to that of the IAU except for the fact that China was previously not a member of the IUB, seemed to go on. However, recognizing China's proposal as a constructive one, Slater enquired *"what were the possibilities of direct negotiations between representatives of Taiwan and the People's Republic of China on this matter,"* to which the Chinese representative replied: *"the President of the Olympic Games Committee had invited representatives from the two sides to discuss with him the question of a single representation in the forthcoming Olympic Games[22] and suggested that perhaps the Executive Committee of IUB could take a similar initiative."* Again a similarity between scientific unions and the Olympic Games!

Slater followed suit, taking into account that the International Congress of Biochemistry would take place in Toronto on the following July:[23] for Chinese and Taiwanese biochemists to attend together the meeting, the Chinese just

> requested that, on the name badges and in other places the residence of the participant is mentioned (in the programme, for example), no distinction be made between Taiwan and the People's Republic of China. They mentioned that, at a recent meeting in Calgary, Canada, *participants were designated as coming from Peking, China or Taipei, China.* This is acceptable to them. Alternatively, they would be perfectly happy if only the town were designated.

(The italics are mine.) These suggestions were unanimously accepted by the IUB; although the suggestion which I underlined seems of a minor nature, it was in fact a breakthrough since it opened the road to the currently adopted scheme for the IAU (except for the fact that "Peking" would be replaced by "Nanjing") and for other Unions.

Almost simultaneously, another development proved decisive for the IAU. On April 16, Patrick Wayman, the Assistant General Secretary, flew on a "private" trip to China (i.e., not organized by the IAU) (Fig. 7.7). Actually, he had been invited as Director of the Dunsink Observatory, near Dublin. Of course, this represented a unique diplomatic opportunity to meet with Chinese astronomers to informally discuss the return of China to the IAU—and also perhaps to make a stopover in Taipei on his way back to meet with their Taiwanese colleagues. A meeting was thus organized in the Peking Hotel, Peking, on April 23, between a Chinese group (Wang Shouguan, now well known, and his colleague Huang Sce Yi [sic], both from Peking Observatory, and Fang Chun, also an acquaintance of us, from the Bureau of Foreign Affairs), and P. Wayman on the other. The main conclusions of Wayman's four-page "memorandum" were as follows:

---

[22]The 1980 Olympic Games (Winter: Lake Placid, NY, USA; Summer: Moscow).

[23]An extraordinary coincidence, since the IAU General Assembly would take place in nearby Montreal less than one month later.

Fig. 7.7 Patrick Wayman and wife Mavis visiting the Great Wall in April 1979 (from "*Météore*," the daily newspaper of the Montreal General Assembly, Issue #3, August 16, 1979; © IAU)

"Professor Blaauw's first suggestion, (a), has considerable merit"; following it would lead to a preferred procedure:

1. The IAU Officers[24] should suggest to Peking and to Taipei that a meeting be held as soon as possible between representatives of the two Academies. (. . .)
2. If such a meeting were held and be successful, an appropriate Executive Committee motion would be added to the Montreal Agenda under By-Law 8(b);[25] (. . .)
3. The Academia Sinica would place a conditional application for full IAU membership, including a category for financial contribution,[26] nominating a National Committee as adhering organization for China, in the form agreed as under 1 above. (. . .)
4. If the meeting referred to in 1 does not take place, or if it results in no agreement between the Academies, the IAU Executive Committee should nevertheless endeavor to prepare a motion for late submission (By-Law 8b) to the Montreal General Assembly—this motion having received the prior approval of Academia Sinica, Peking, and, if possible, of the

---

[24]IAU jargon. . . The "Officers" are the President, the President-Elect, the General Secretary, and the Assistant General Secretary.

[25]Referring here to the IAU Statutes of 1970; By-Law 8(b): "A motion or proposal concerning the administration or budget of the Union which does not appear on the agenda prepared by the Executive Committee, or any amendment to a motion that appears on the agenda, shall only be discussed with the prior approval of at least two thirds of the votes of Adhering Countries represented at the General Assembly and having the right to vote."

[26]On February 22, 1980, P. Wayman would write to Y.C. Chang indicating "Category 4 with 6 units of contribution, so that the dues of your organization to the Union for 1980 amount to 8790,-- Swiss Francs." After further correspondence with Chang, the dues were reduced to Category 3 and 4 units = 5860 Swiss Francs. This is equivalent to 2578 € (2021) (https://fxtop. com/fr/historique-taux-change.php).

Academy of Taipei. *This motion would probably define the manner in which China should adhere to the Union. (See Note B below.)* [The italics are mine.]
5. The conditional application of Academia Sinica, Peking, for IAU membership would then be similar to that referred to in 3 above (. . .)
6. If the IAU Executive Committee were unable to follow the course outlined in 1–5 above, the Academia Sinica would be prepared to consider Professor Blaauw's initiative (c) (. . .) [27]

And Wayman gives the details in its Note B:

Two possible formulations for the mode of adherence of China to the Union would, under conditions of 4 above, be:

(i) Through the National Committee for Astronomy of the Chinese Astronomical Society, Scientific and Technical Association, Peking, China, with the conjoined Committee for Astronomy of the Astronomical Society of Taiwan.
(ii) Through a dual adherence provided by

   a. The National Committee . . . . Peking, China, and
   b. The National Committee for Astronomy of the Astronomical Society of Taiwan, China.

In Note D of the minutes, Wayman however urges diplomatic caution: *"The organisations represented at the meeting, namely the Academia Sinica, Peking, and the Executive Committee of the IAU, do not consider themselves bound to follow the proposals herein made."*

Still missing, however, was the reaction of Taiwan to Blaauw's letter of April 9. It came not from the Academy of Sciences but from the President of the "Astronomy Union of the Republic of China," C.S. Shen (or Shen Junshan in pinyin, 沈君山, whom we already mentioned and whom we shall meet again later),[28] writing on May 1:

We do not claim to represent the scientists in Mainland China. (. . .) We do not object to admitting PRC's Astronomical Union into I.A.U. provided they do not claim to represent the scientists here in Taiwan. (. . .)

   The eventual solution by the IAU on the China 'issue' [sic] should, I hope, adhere to the principle of open and universal participation, and reflect the reality that Taiwan and Mainland China are two independent scientific communities which at present can only pursue their development separately, thus, they cannot be realistically represented by a single delegation at the time being.

As Blaauw had anticipated, Taiwan was not playing along: Shen did not reply to Blaauw in the way he had hoped, i.e., commenting on his proposals and expressing his opinion about the three suggestions ("a" to "c"). Nonetheless, there was indirect

---

[27] In this document, "Academia Sinica" means "Peking," which is confusing because it was (and still is) the official name of the Academy of Sciences of Taiwan (see also Footnote 3 in Chap. 4). Elsewhere in the same document, this Academy is called "Taiwan's Academy" by P. Wayman.

[28] C.S. Shen had a PhD (U. Maryland, USA) in spectroscopy. Returning to Taiwan in 1973, he became Head of the Faculty of Sciences of the National Tsing Hua University (國立清華大學) in Hsinchu. Eventually, Shen became President of this university in 1993. See also the additional information provided by Fu&Ye below.

acknowledgment that Taiwan "recognized" the Astronomical Union of China (here "Mainland China," since Taiwan still named itself the "Republic of China"), and indirectly expressing a preference for a "reciprocity" in treatment by the IAU, namely two equal representations, or, in the words of Wayman's Note B which he hadn't seen yet, a *"dual adherence."*

Therefore, willy-nilly the idea of a "dual adherence" for the same "country" took shape. At this point, one cannot but compare it with the scheme initially proposed by Sadler and Oort at the Herstmonceux meeting of 1959 (see Sect. 3.6): removing all reference to two separate countries ("two Chinas"), but using only "China," this scheme announces Wayman's "dual adherence" 20 years later. The similarity is even more striking if one considers Oort's reply to Y.C. Chang on December 2, 1959 (quoted in Sect. 3.8), explaining the scheme "to consider China's representation in the IAU as *'a temporary, dual one, with one adhering organization located in Peking, and the other in Tapei',"* and is even visionary in that the correct geographical reference to the Chinese Astronomical Society, i.e., Nanking (Nanjing), is given in the scheme, rather than Peking!

Unfortunately, some time was wasted after these efforts, because neither Wayman's visit to Taiwan could be arranged (the Taiwanese had not finalized the answer to Blaauw in time) nor could the joint China–Taiwan–IAU planned meeting in Tokyo take place (the relevant abundant correspondence in the Archives shows that in fact there was also to some degree a fear to move too hastily). But the spirit was there, and the perspective of the IUB Congress in Toronto was showing the way. On June 14, C.S. Shen wrote to Blaauw:

> I have participated, mostly as an observer, in the IUB meeting this afternoon,[29] where the delegation of R.O.C. discussed the China issue with Dr. Whelan [IUB General Secretary] and Dr. Slater. As you know very well the problem in I.U.B. is very much similar to that of I.A.U. Dr. Slater must have already informed you of the proceedings and results of that meeting. (. . .) I would not rule out the possibility of delegates from the two sides exchanging views about the China problem at Montreal one or two days prior to the General Assembly. One does not have to make it very official at the beginning. That approach will save lots of trouble such as protocol problems and other political considerations.

Several exchanges took place, first by mail, then soon by "cables" (telegrams), as time was running out. But the winds were this time blowing the clouds away. On July 4, Chien Shih-liang (Taiwan) sent a cable to A. Blaauw (c/o E. Müller): *"(. . .) we agree to your proposed formulation to achieve participation of all chinese astronomy in iau quote dual membership of china through the adhering organizations astronomical society of china—peking and astronomy union of china—taipei unquote. (. . .)."* Figure 7.8 shows the cable as it arrived in Geneva.

On July 24 (the General Assembly was now just three weeks away), Blaauw circulated a cable to all Executive Committee members: *"regarding participation china in iau this is to inform you that both parties have now favourably reacted to my*

---

[29] Because, coincidentally, it took place at C.S. Shen's university in Taiwan.

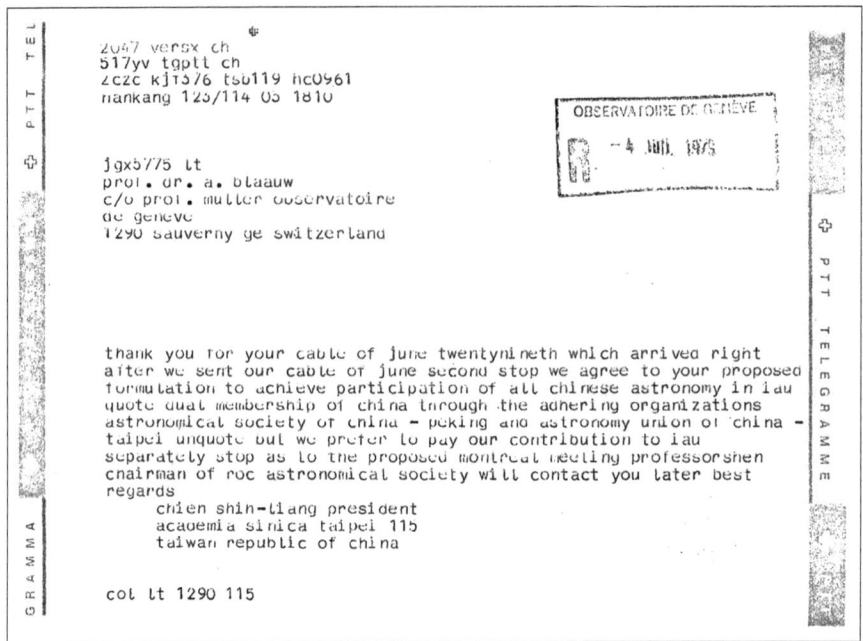

Fig. 7.8 Cable sent by the President of the Academia Sinica in Taipei on July 4, 1979, informing Blaauw of his agreement to the "China-Peking" and "Astronomy union of China-Taipei" formulation for the adhering bodies of China and Taiwan, respectively (IAU Archives)

invitation for continuation discussion prior to general assembly stop china peking delegation will consist of six people. looking forward to meeting you in montreal."

The caption to Fig. 7.9 lists the members of the Chinese delegation, headed by Y. C. Chang (i.e., Zhang Yuzhe). Of the nine IAU individual members of Taiwan at the time (see Table 3.1 in Sect. 3.6), Shen (Fig. 7.10) would be its sole representative.

The momentum was accelerating. On July 28, the IUB Circular n° 128 *"To the Adhering Bodies and the ICSU Family"* states: *"The following agreement was reached, as a solution to the representation in IUB of biochemists from China. This agreement, and the steps to be taken to implement the agreement, were presented to the 10th General Assembly meeting* [in Toronto] *on July 11."*

The main points of the Agreements were:

1) *For the time being there will be two Adhering Bodies*
2) *(. . .)*
3) *The membership of these Adhering Bodies in IUB shall be designated by their being listed in By-Law l as:*

China The Chinese Biochemical Society
The Biochemical Society of Taipei, China

Fig. 7.9 The Chinese delegation at the Montreal General Assembly. *Left to right:* Zhu Jinning (Staff, Bureau of Foreign Affairs in Peking, 朱进宁), Ye Shuhua (叶叔华), Hong Siyi (洪斯溢), Zhang Yuzhe (张钰哲), Zhao Xianzi (赵先孜), and Yi Zhaohua (易照华) (From Fu and Ye [2009], original Chinese edition)

Fig. 7.10 Shen Junshan (C.S. Shen), the Taiwan delegate. His obituary (in Chinese) can be found at: http://alumni.site.nthu.edu.tw/p/404-1346-147042.php?Lang=zh-tw (See text for details; credit: National Tsing Hua University's Alumni Association, Taiwan.)

(The italics are in the original.) The IAU archive also keeps a record of a detailed article by Joan Hollobon,[30] a medical reporter for the Toronto *Globe and Mail,* dated July 13 and entitled: *"China to join scientists' group which includes Taiwan."* It opens with the sentence: *"For the first time since the People's Republic of China was formed 30 years ago, scientists from the Mainland have agreed to join an international society which includes a member group from Taiwan."* Interviewing W.J. Whelan, the UIB General Secretary: "[he] *hopes the achievement will be a breakthrough that will enable China to join many other scientific bodies."*

Indeed, the Union next in line was the IAU, so let's move on to what happened at the Montreal General Assembly, which opened three weeks later, on August 14, 1979.

By a strange coincidence, when Blaauw and Zhang Yuzhe met, they realized that they had been together, even sitting next to each other on the group photo (along with Otto Struve and Jan Oort) some 32 years before, in 1947, during a stay at Yerkes Observatory near Chicago![31] (Fig. 7.11).

As is well known, A. Blaauw himself tells the story in his book (pp. 201–204), so it is worthwhile to have it told here for the first time from the Chinese side by Fu&Ye, which is in a way another first-hand testimony since Ye Shuhua, a radioastronomer educated in the USSR and future IAU Vice-President (Fig. 7.12), was herself a member of the Chinese delegation (see above, Sect. 7.1, and Fig. 7.9; in Appendix C by Pierre Léna, we learn that during the Cultural revolution she had been protected by Zhou Enlai as a gardener in the old Zi-Ka-Wei Observatory, near Shanghai).

Here are some large excerpts from Fu&Ye:

From the Mainland's side, the head of the delegation was Zhang Yuzhe, who was the long-time President of the Chinese Astronomical Society. His deputy was Zhao Xianzi (1926–1996), who was from the Purple Mountain Observatory. There were also four other members that went to Montreal in August with them. They included Ye Shuhua (1927–) who worked at the Shanghai Observatory, Hong Siyi of the Beijing Observatory, Yi Zhaohua of the Nanjing University (1932–2017) and Zhu Jinning who was an official from the China Association for Science and Technology. On the Taiwan side, Shen Junshan (1932–2018) was the only delegate. He was the then President of the Taiwan Astronomical Society. The two sides started their talks from August 13th till 24th, which lasted longer than the Assembly agenda itself. It was a friendly but laborious discussion. The Mainland side wore a card which was written 'China: Nanjing', and Shen Junshan wore a card which stated 'China: Taipei' on his chest as well. Since the Chinese Mainland was yet to return to the IAU, its delegation did not participate in the General Assembly, but took part in the sideline academic discussions. They did so to show off the strength of the Mainland's astronomical field on one side, but also to prevent possible misunderstandings at the Assembly on the other side.

The sole Taiwan representative to the Montreal Assembly, Shen Junshan, was born in Nanjing; his paternal family roots were in Yuyao, Zhejiang. (. . .) Since his father occupied prominent role in the ROC government in Taiwan, he and three other youngsters at the time were given the nickname of 'Four Princelings of the Chinese Nationalist Party' between

---

[30] https://www.nasw.org/article/joan-hollobon-named-officer-order-canada
[31] Zhou (2022).

Fig. 7.11 Yerkes Observatory staff, November 1947. *From left (back row, standing)*: Daniel E. Harris, Guido Münch, Henry Chun, Su-Shu Huang, Douglas Duke, Mrs. Marshall H. Wrubel, Marshall H. Wrubel, Mr. Clearman, Arne Slettebak, William Bidelman, Roy Wickham, Irene Hansen, Frances H. Breen, Marjorie Hall Harrison, Gertrude Peterson, Robert H. Hardie, Arthur D. Code, K. N. Rao, Henry G. Horak, Marvin L. White, Frank N. Edmonds, Anne B. Underhill, Nancy G. Roman, Mr. Robinson, John Vosatka, Margaret Phillips, John G. Phillips, Mr. Sinha. *(Front row)*: Thornton L. Page, William W. Morgan, Charles Ridell, *Yu-Che Chang*, *Adriaan Blaauw* (IAU President, 1976–1979), Luise Oettinger Herzberg, Gerard P. Kuiper (IAU President, 1958–1961), George Van Biesbroeck, William A. Hiltner, *Otto Struve* (IAU President, 1952–1955 and Yerkes Observatory Director, 1932–1947), Gerhard Herzberg, Marguerite Van Biesbroeck, Fred Pearson, Subrahmanyan Chandrasekhar. *(Credit: The University of Chicago Photographic Archive* [apf6-00496], *Hanna Holborn Gray Special Collections Research Center, University of Chicago Library)*. Major protagonists of the China-IAU divorce and reconciliation were there and are indicated (in italics). G. Van Biesbrock had been Y.C. Chang (Zhang Yuzhe) PhD thesis advisor in the 20s, and S. Chandrasekhar (Nobel laureate 1983) was a good friend of him

Fig. 7.12 Ye Shuhua lecturing on VLBI (Very Long Baseline Interferometry), with a map of China in the background. VLBA is a technique in radioastronomy to study objects appearing very small, or having very small structures, on the sky, combining radiotelescopes located at large distances from one another. Photo taken around 1980, when the VLBI project was about to be launched (Credit: From an interview of *Ye Shuhua at 92*: Shanghai Observer 上观, November 19, 2020, in Chinese)

1970s–1990s. Shen Junshan was not only a learned scholar in astronomy, but also a cultured fellow in Classical Chinese arts. He was a renaissance man. His special talents made him a perfect candidate to manage the political debate in Taiwan, as well as handling the Cross-Strait relations. *In October 1979, the agreement between both sides in regards to their adhesions to the International Olympics Committee was a result of Shen Junshan's participation* [see above; the italics are mine]. He represented Taiwan as a member of the ROC Olympics Committee and discussed this matter with the PRC Olympics Committee officials in Nagoya in Japan. From this meeting, the Mainland and Taiwan delegation agreed to Beijing's participation in the International Olympic Committee. And Taiwan would use the title 'Chinese Taipei' to continue to participate in Olympic events. *This China and 'Chinese Taipei' formula was kept as a model in solving the Taiwan question for many international organizations to come.* After the Montreal Assembly, Shen even visited the Mainland on multiple occasions, he contributed greatly for the promotion of Cross-Strait academic cooperation and developments.

The most challenging obstacle was both sides' political positions. However, after a series of sincere exchanges, they all welcomed each other in participating the IAU activities. (. . .) Shen Junshan treated the Mainland delegation with politeness and respect, he called Zhang Yuzhe as his 'elder'. Their talks included many aspects and were really clear. No matter what the topic was, they were all under the principle that China was indivisible. Any 'Two Chinas', 'One China, One Taiwan' suggestions were not to be permitted. Near the end of their talks, the two sides focused on the conflict of finding a suitable name for Taiwan's delegation that would be not only in line with the above-mentioned principles, and a name which could be agreed upon by both sides. The Mainland suggested names such as 'Chinese Taiwan Astronomical Society', or 'Chinese Taiwan Astronomy', but they were all rejected by Shen Junshan. This deadlock frustrated Blaauw and he kept on walking back and forth in his office, while trying to find a better name for the Taiwan delegation. He even sighed that 'this was totally incomprehensible, these ideas which seemed to be good yet were still rejected!' Blaauw's frustration might be a result of Western-Eastern cultural differences. Shen Junshan who was by himself was also really preoccupied due to lack of help. He called Taipei nearly each day to forward the talk contents and ask for permissions. Even so, the three sides at the talk table could not find a suitable name for Taiwan to remain at the IAU that was accepted by everyone.

In the end, the Mainland delegation suggested to temporarily suspend Taiwan delegation's name debate. This was to be treated in later times. At the beginning, this proposal was rejected by the Executive Committee of the IAU. But as the Assembly was reaching its end, the three sides realized quickly that if they could not reach a consensus through these talks, the Taiwan question would still remain after Montreal. Based on this understanding and the PRC delegation's relentless efforts, the Executive Committee finally adopted the temporal suspension of the Taiwan delegation's name question in the end. Since this proposal was also not yet approved by the Taiwanese side, the Executive Committee further adopted a compromising strategy, where the agreement was not to put forward for a vote at the General Assembly, but to pass by a letter exchange. In doing so, it avoided the possibility of lack of support in the Assembly's voting process. This truly particular agreement was called the Montreal Agreement.

These letters are given in full in Fu&Ye (Appendix D), and as signed originals from the IAU archive in Appendix B.5, together with the draft resolution that was submitted to the General Assembly. This resolution was not the final resolution that Blaauw originally had in mind, but it did authorize the Executive Committee to finalize the negotiations with Taiwan so as to have the full agreement ratified by the next General Assembly. In Montreal, the venue had not yet been decided (there were three proposals), but the ratification did eventually take place in Patras, Greece, in 1982. All these documents can be found in printed form in the IAU Transactions Vol. XVII B (1980) and Transactions Vol. XVIII B (1983).

In his opening address in Montreal, Blaauw had declared:

I wish to say a few words on an item of special importance; that item is the restoration of full participation of Chinese astronomers in the Union. The absence of astronomers from the Mainland of China at our General Assemblies since 1961 has been a matter of great concern for members of the Union, and therefore also of the Executive Committee; and that is particularly true for the present Executive Committee and that of the preceding period. Many of you will remember the words spoken with regard to this problem by the previous President of the IAU, Professor Leo Goldberg, at the Grenoble Assembly. To me, and to my fellow officers, it seemed one of the most urgent problems to take up when we assumed our duties three years ago.

(. . .)

We have been very fortunate in having had with us here in Montreal, for almost a week now, delegations of the astronomical communities of both the Mainland of China and Taiwan, so as to approach the solution further, following the preparatory discussions and correspondence; in these preceding activities also, many astronomers outside the officers of the Union have participated for which we are most grateful.

(. . .)

Not only will we become much better aware of the progress of astronomy on the Mainland of China and, at last, be in a better position to share with them the progress in astronomy in other parts of the world, *but I believe this experience should also strengthen our awareness of the rewarding, albeit modest, role we scientists, and particularly we astronomers, can play in bridging the gaps that unfortunately tend to keep this world divided.*

Blaauw then concluded: *"I sincerely hope that by the time of the conclusion of this General Assembly, I may be in a position to inform you in more detail on the arrangements that we have been been able to make in this matter."*

The goal was essentially reached. P. Wayman, the incoming General Secretary, said in turn during the closing ceremony:

> I wish particularly to express admiration, in my recent experience, to the work of our retiring President, which I have witnessed at close hand, in achieving a triumph in the diplomatic fringe of our activities. If the solution is not complete, that is certainly not due to any negligence on his part. His patience in this task has been unbelievable and I do not see how any more could have been achieved.

It took indeed a couple of months, and a lot of discussions on important details (including on the "category," i.e., the level of contributions to the IAU, the list of individual members from China, etc.), for a final solution to be agreed by all parties, and to be ratified "by acclamations" in Patras (Fig. 7.13).

Fig. 7.13 Members of the The Chinese delegation at the Patras General Assembly, together with other Chinese participants based in Europe. *From left to right:* 何香涛 He Xiangtao, 李正心 Li Zhengxin, 王绶琯 Wang Shouguan (Head of delegation), 曹昌新 Cao Changxin, 曲钦岳 Qu Qinyue (delegate), 叶叔华 Ye Shuhua (delegate), 邢骏 Xing Jun (delegate), 须同祺 Xu Tongqi , 张家祥 Zhang Jiaxiang (delegate), 童傅 Tong Fu (delegate), 赵君亮 Zhao Junliang (From Fu and Ye [2009], original Chinese edition)

7.4   A Final, General Agreement: The IAU, ICSU, and Other Unions

Interestingly, while there was a lot of correspondence between the IAU and China (Y.C. Chang) on the one hand, and between the IAU and Taiwan (C.S. Shen) on the other, and in view of the fact that Taiwan or China could not agree on early suggestions about Taiwan's "name," the recourse to common solutions, adopted on similar grounds by other Unions, was mentioned more and more often.

For instance, in a letter to Wayman (now the new General Secretary) on December 10, 1979, Y.C. Chang wrote:

> (...) In order to propitiate settlement of the question of the representation of China in IAU, we are willing to make another concession and agree to your proposal that 'Astronomy Union located in Taipei, China' as the title of Taiwan astronomical organization. But Dr. Shen demanded that this title be preceded by the word 'Chinese'. In so doing, the nature of the organization becomes ambiguous. We can't agree on this point.
>
> The name we suggested [in Montreal: 'Chinese Taiwan Astronomical Society'] is after the pattern of IUPAC and IUB. The resolution of IUPAC and IUB has been accepted by Taiwan authority. (...) *Following the precedent of the International Olympic Committee, IAU should make a decisive resolution.* (The italics are mine.)

Indeed, in the preceding section, we have seen the "dual adherence" agreement that was reached in Toronto by the IUB just before the IAU General Assembly in Montreal. Such a "quick" agreement was made possible because, as we have already mentioned, the IUB, like all other major Unions, is composed of "national members," "committee members," etc. having some kind of mandate. But the case of the IAU was more delicate, since it is essentially composed of individual members (over 4500 in Montreal, three times the number at the time of the Berkeley General Assembly of 1961, in addition to the national members: 48 in Montreal, from 35 in Berkeley), and therefore as an organization it is more sensitive to how a given wording might affect one's personal perception...

Eventually, after many "triangular" exchanges, the IUB type wording prevailed. On January 25, 1980, Wayman proposed that China would be represented by two member organizations, respectively named:

– Chinese Astronomical Society (Nanking)
– Astronomy Union located in Taipei, China,

to which Y.C. Chang agreed on March 12.

Translated into the countries adhering to the IAU, i.e., in the listing to be printed in the forthcoming Transactions Vol. XVII B (to appear in late 1980), Wayman finally informed ICSU on May 12, 1980, of the official decision:

-	China: Chinese Astronomical Society Purple Mountain Observatory, Nanking, China	(year of adherence: 1935)
-	China: Astronomy Union located at Taipei, China Taipei Observatory, Taiwan, China	(year of adherence: 1959)

The abbreviated names adopted, when necessary (for instance on name tags, ballots, etc.), were respectively: "*China, Nanking*" (Nanjing) and "*China, Taipei.*" The "Taipei Observatory" was later replaced by "Academia Sinica" after some internal reorganization in Taiwan.[32]

Attention, however, was given to other details, sometimes hiding unnoticed the same important issue, that of defining a "country." For instance, in a letter received on July 2, 1982, Y.C. Chang (now signing "Zhang Yuzhe," in the recently adopted pinyin spelling) points out that

> it is important to note that in this list [i.e., above] whenever 'Adhering Country' is mentioned it should mean 'China' and neither of the two adhering bodies should be referred to as 'Adhering Country'—they should only be called 'Adhering Bodies'.

This remark was accepted and led to a correction to the Statutes, insofar as the terms "Adhering bodies" and "Adhering country" indeed had tended to be used more or less indifferently throughout the text.

As for other Unions, and quite logically, much the same agreements were reached. On September 8, 1979, right after the end of the Montreal General Assembly, the chemists of IUPAC decided that

> "For the time being there will be two organizations" and "The membership list of Adhering Organizations will read as follows:
>
> - China The Chinese Chemical Society
> -   The Chemical Society Located in Taipei, China"

Other Unions had the same agreement later. IUPAP, for instance, did it when the Chinese Physical Society was admitted in 1984: Taiwan, as the "Republic of China," had been a member before but for some reason China didn't join earlier. In the current IUPAP listing, both the "Chinese Physical Society" and the "Physical Society located in Taipei" appear, but mentioning 1984 as their common year of adherence.

ICSU itself reached the same agreement, with CAST in Beijing and the "Academy of Sciences located in Taipei" in Taiwan, but this time the year of adherence is given as 1937 (the original year for China) for both. ICSU organized its 22th General Assembly in Beijing as early as 1988 (the IAU did the same only in 2012, but its first Chinese Vice-President, Ye Shuhua, was also elected in 1988).

More recently, this type of agreement was Generalized to other locations: For instance, IUTAM (*International Union of Theoretical and Applied Mathematics*) currently has *three* adhering bodies: "China" (= China-Beijing) and "China-Taipei," with 1980 as common year of adherence, to which "China-Hong Kong" was added in 1996. A very flexible formula indeed...

---

[32] For linguistic details on how subtle differences were expressed (and agreed) in Chinese but could not be easily translated (and accepted) into English, hence in the official texts, see Wang, S.-Y. et al. (2022), and Shen (2004).

And what about IUGS and IUGG, the dissidents which had expelled Taiwan in 1976 and 1977, respectively, in contradiction with ICSU policy?[33] Looking at their current websites, IUGS, which was founded in 1961 with support from UNESCO, currently has its Secretariat in Beijing, "supported by the Government and China and the IUGS".[34] This is yet different: IUGS lists its Chinese Adhering members as "China, People's Republic of" and "Taiwan, China," but their websites are in Chinese only (their page in English is inactive) and we could not retrieve their years of adherence. As for IUGG (one of the first Unions created in 1919), it appears that for some reason it re-admitted Taiwan very late (in 1995).

In spite of these differences, the overall conclusion is clear: From the beginning to the end of the crisis, the IAU was far from being alone in its admittance of Taiwan, the divorce from China, the subsequent 20 years separation, "crossing the desert," and their quick reconciliation starting from 1979. But for all Unions, the cost is a still delicate balance to comply with the will of China to be "one country," on one side, and to safeguard the activity of Taiwanese scientists, on the other.

---

[33] Chapter 6.

[34] See https://www.iugs.org/history

Chapter 8
Conclusions

8.1 The Admission of Taiwan: Was It "Political"?

At the time of the Grenoble General Assembly in 1973, the article in "*La Gazette d'Uranie*," although signed by only six astronomers, probably reflected the opinion of a larger number of participants, especially among the more senior ones that had already attended the Berkeley General Assembly 15 years before. I repeat here the main criticism:

> We believe that the decision by the IAU in 1958 to admit 'The Republic of China' of Taiwan to membership was a mistake. It is now Generally admitted that the application of Taiwan for membership was politically motivated. It was part of a plan by the US State Department under Dulles to gain Taiwan's admission to as many international organizations as possible, thereby isolating the Peoples Republic of China internationally. (. . .) we believe that the IAU unwittingly allowed itself to be used for political ends.

I (TM) had been myself always disturbed by this statement, which was contrary to my personal experience of the way the IAU Executive Committee worked. While adopting a rather neutral position on the subject of the admission of Taiwan, Blaauw's book does convey the idea that, at least initially, the US State Department directly pressured the IAU to admit Taiwan in exchange of accepting to grant visas to participants from the Mainland China (with whom there were no diplomatic relations). This view is the one held by Fu&Ye, and by my colleagues G. Miley and Xiaowei Liu.[1]

But after the extensive search done for the present chapter, in the IAU archives and also about ICSU and other Unions, I'm coming to a significantly different conclusion.

First, since we're here referring to politics, one cannot avoid considering the international situation at the time. In short: The world was on the brink of a new World War. In 1958, the year of the IAU Xth General Assembly in Moscow, but also

---

[1] Miley (2022) and Liu (2022).

© The Author(s), under exclusive license to Springer Nature Switzerland AG 2022
T. Montmerle, Y. Zhou, *China and the International Astronomical Union*, Historical
& Cultural Astronomy, https://doi.org/10.1007/978-3-031-01787-2_8

that of the 8th ICSU General Assembly in Washington, the Korean War, in which there had been bloody fighting between the USA (leading a UN coalition) and China (reinforcing the North Korean troops), and even the USSR (in the air), had ended only five years before. Armed confrontations between China and Taiwan, backed by the USA, took place in the Taiwan Strait that same year. In 1961, on the very weekend preceding the Berkeley General Assembly, the Berlin Wall crisis erupted, with a potentially dangerous face-to-face escalation between the USA and the USSR.

As we have seen in Sect. 3.2, based on D. Sadler's testimony in his 1976 "personal account" (Appendix A), when he and J. Oort attended the ICSU General Assembly in Washington, they witnessed that Taiwan, under ICSU guidance, was applying to several Unions at the same time. We can only guess that UNESCO, a major partner of ICSU, played a role here, as a UN organization of which Taiwan was a member, and of course, as the Korean War had demonstrated, in which the USA were highly influential.

In this context, the direct involvement of the State Department into the early stages of the admission of Taiwan by the IAU (which cannot be denied) is *purely circumstantial*: It only added drama to the *independent issue* of the Executive Committee decision. The "blackmail" mentioned by Sadler played a role only because the General Assembly following Moscow was to be held in the USA, and because there was a critical visa issue since IAU General Assemblies are a large gathering of individual members, not of representatives of National members and of various committees like the other Unions. The situation would have been totally different if, for example, the General Assembly after Moscow had been held in Hamburg (venue of the 1964 General Assembly): In that case, it is clear that the State Department, as such, would not even have appeared in the story at all. In any case, *there was never a pressure on the IAU to expel China* (contrary to what would be the case for Taiwan by UNESCO later): *China withdrew on her own will only for internal political reasons,* and apparently from the story of the Olympic Games, this China–Taiwan "incompatibility" in international organizations was decided only in 1958, when the Taiwan Strait crisis erupted, and on the eve of the disastrous Great Leap forward.

Therefore, in my view, the proof so frequently mentioned of the fact that politics intervened in the Executive Committee decision to admit Taiwan under the pressure of the State Department, actually misses the point: As the case of ICSU and of the other Unions showed, *the admission of Taiwan was a general decision,* even if in all cases it resulted in the unavoidable withdrawal of China.

8.2   Was It Legal?

So let's forget about the State Department, as did the members of the Executive Committee at the Herstmonceux meeting in 1959. Before even considering whether it was appropriate at all to consider Taiwan's application, it was necessary to verify that it would be "legal," i.e., in conformity with the Statutes, as they were in force in 1958 (and since 1932).

The *"Gazette d'Uranie"* article mentioned the argument that *"there was a negligible amount of astronomical activity* [in Taiwan]," which is a question of Statutes, not of politics. First of all, the Statutes said nothing about the "astronomical activity" requirements, and even though this argument was one of those used by Kukarkin and Šternbek in their protest to the Executive Committee and in the resolution submitted to the Berkeley General Assembly (see Chap. 3), it can be easily dismissed: There was already at that time ample evidence that the IAU had adopted a fully inclusive policy and accepted the adherence of countries even with small astronomical activity (see Table 2.1, Sect. 2.1). Not only was this justified in the aftermath of the destructions of WWII, but also in view of the emergence of developing countries, as recalled by Goldberg in his opening address at the Grenoble General Assembly. And the compelling evidence was the admission of the Democratic People's Republic of Korea (aka North Korea) at the Berkeley General Assembly, i.e., at the same time as Taiwan.

Actually, there was indeed "some astronomical activity" in Taiwan at that time, even though it may have had a low visibility abroad. It followed the progressive introduction of modern astronomy by the Japanese,[2] which continued until the 1941 total solar eclipse (the same one that Zhang Yuzhe observed in Langzhou),[3] just a few weeks before Pearl Harbor and the entry of the USA in the war on the Chinese front. Taiwan was liberated in 1945 by Chiang Kai-shek and his Nationalist army, but after 1949 the island fell under his dictatorial regime, and life continued to be difficult for the few scientists or professors involved in astronomy, including those that had decided not to follow their colleagues but to leave the Mainland. Yet, an Astronomical Society did emerge, and in June 1958 a 49-member strong "Astronomical Society of the Republic of China" held a "reorganization meeting" in Taipei, as Taiwan's application to the IAU was following a tortuous path to the Executive Committee in Moscow (Fig. 8.1).

More to the point was the fact that, as demonstrated in Sect. 3.5, the Statutes were fully adapted to accommodate the astronomers of *both* Taiwan and China: Taiwan through its membership with ICSU (Statutes 3(a)), and China through its affiliation with a National Committee of Astronomy recognized by the Executive Committee of the Union (Statutes 3), which had been its status all along since its adherence to the IAU in 1935.

Thus, the conformity with the 1958 Statutes is well established. In addition, the 1961 General Assembly confirmed the Executive Committee decision by a majority vote of the National Members present, a move which was not a statutory obligation. Admittedly, the Executive Committee reacted negatively to a resolution proposed by two of its members in protest to the admission of Taiwan, but given the importance of the issue, it is likely that the Executive Committee, on the advice of D. Sadler,

---

[2] Wang et al. (2022).

[3] Zhou (2022).

Fig. 8.1 The memorial picture of the reorganization meeting of the Astronomical Society of the Republic of China in Taiwan in June 1958. The first President is [8] Bing-Ran Jiang. The important members include [1] Kenneth T. C. Cheng, [5] Ping-Tse Kao, [11] Mo Tsao, [13] Jie-Chien Huang, [23] Shao-Chin Chang, [38] Chang-Hsien Tsai, [41] Li-Chen Wang (Courtesy S.-Y. Wang, with permission of the Taipei Astronomical Museum)

would have submitted sooner or later the admission of Taiwan to a ratification vote of any General Assembly. As noted previously, this procedure would become mandatory when the Statutes were revised in 1970 (Brighton General Assembly), i.e., many years later. And last but not least, there was no crisis within the Executive Committee after Berkeley: Nobody resigned, Kukarkin went through his second mandate as Vice-President without ever mentioning Taiwan again, the Soviet astronomer V. Ambartsumian succeeded Oort (who remained in the Executive Committee as Advisor), Sadler accomplished a second term as General Secretary, etc.: The decision to admit Taiwan was taken, there was no turning back.

Now the same group that wrote the Grenoble article continued their action after the General Assembly, and "a colleague from Leiden"[4] submitted to Blaauw in September 1976 a more elaborate document: The membership of Taiwan had

---

[4]Blaauw's expression in his book (p. 197): G. Miley, disclosing himself in Miley (2022).

become "invalid" because it didn't conform with the revised Statutes, adopted in 1970, which used the term "countries" more strictly than previously, dropping the sentence "any dependency in which independent astronomical activity has been developed." And Taiwan could not be a "country," since China claimed that "Taiwan was a part of China"; in other words, the argument went, Taiwan now had to be expelled. As told by himself in his book (p. 197), Blaauw *submitted* [the document] *to a legal advisor whose counsel I had learned to appreciate during my Directorship of the European Southern Observatory* [ESO]. *He confirmed that, contrary to the submitted document, the IAU Statutes offered no grounds for expelling Taiwan.*

In reality, it was likely that Blaauw contacted the ESO advisor as a neutral person simply to avoid a potential internal conflict within the Leiden Observatory, because he could have himself provided very easily the answer: In the revised Statutes, only one condition is given (Art. 7): *"The adherence of a country (...) terminates if the country withdraws from the Union."*

Sadler, in his "Personal account" of 1976 (later: in December), *"call[s] attention to the explanatory note to Statute 7"* (see, for example, Information Bulletin No 23, page 26): *"Once a country has become a member of the Union, there is no plausible ground for expelling it."*[5]

It may be worth going here one step further and quote Sadler's further thoughts about this note, especially having in mind the contemporary context of the UK:

I do not entirely agree with this statement (due mainly to Jappel),[6] but I do not think the Union should allow itself to be coerced into expelling a country for reasons *outside* its own concerns.

There are lots of problems. What happens when the U.K. devolution bill becomes law,[7] or if Scotland becomes even more separated? Will the British government be responsible for astronomical activity in Scotland? *The present absence of a definition of a "country" in its Statutes (as adopted at G.A. XIV, 1970) was a deliberate attempt to avoid the impossible task of specification that would involve political considerations.*

[My italics.]

It may well be that, in changed circumstances, the IAU should align itself with ICSU and UNESCO and follow UN definitions, essentially identifying "countries" as the National Members of the UN. This would mean following the UN lead in regard to South Africa, etc., but it would simplify the IAU problem: *Personally* I would be opposed to so devolving decisions to the UN (or UNESCO), but the possibility of a unified policy between the scientific Unions under ICSU should not be excluded.

---

[5] Except in case of defaulting in the payment of dues, as was case of North Korea in 1995, as a result of the collapse of the USSR (Montmerle, 2022a).

[6] Arnost Jappel, the IAU Executive Secretary, was a Czech lawyer (Pecker, 2019).

[7] At the time a project pushed by the Labour party; it involved the "devolution" of some powers from the UK Parliament to local Parliaments (Scotland, Wales and Northern Ireland). It would eventually pass by referendum 20 years later, in 1997, and is still in force.

(In practice, the issue would never be raised: the "China problem" would be solved by China itself three years later.)

In any event, ICSU and all the Unions, having essentially the same Statutes, always invoked them when opposing UNESCO and refusing to expel Taiwan: There is perhaps no better proof that these Statutes were so legally binding that the two Unions having expelled Taiwan had first to revise them internally to allow it.

8.3   Was It Right?

The last question, and perhaps the most relevant of all, is: Given that, forgetting about the State Department, and assuming there could have been ways to circumvent the Statutes (which Sadler would never have allowed anyway), for instance by delaying again and again the admission of Taiwan (thus transferring somewhat cowardly the "hot potato" to the next Executive Committee), was the decision taken in 1959 "right," at the obviously high cost of losing China, perhaps for ever?

Sadler comments: *"After nearly twenty years I still feel that the action taken by the Executive Committee over Taiwan's application was the only correct one—in spite of the general revulsion against its political motivation."* My own view is that, all things being considered, the Executive Committee didn't have much choice. Even though I claim that the State Department interference was circumstantial, it was there. And Taiwan was an ICSU member, and it was or would very soon be a member of other Unions, probably also for high-level geopolitical reasons (links between ICSU and UNESCO?). Therefore, the IAU could hardly have adopted a different position, once the conformity with its own Statutes had been established.

But the China threat to withdraw upon the admission of Taiwan was also a pressure on the IAU, of a different nature, but its constant claim to represent "one China" was (and still is) fundamentally political nonetheless, especially since Taiwan was initially making the same claim given its status of UN member.

So, as discussed at length in Chap. 3, the Executive Committee was caught between opposing forces. The IAU being a non-governmental organization, applying its Statutes was the only way out of this impossible dilemma. In other words, the Executive Committee was not "right" or "wrong": it only strictly, if reluctantly, "applied the law" of the IAU. The correctness of this approach was fully vindicated 20 years later by the very similar, and general, so-called *"dual solution,"* summarized by the short names for the adhering organizations "China, Nanjing" and "China, Taipei."

8.4   Why Did the "Happy Ending" Happen So Quickly?

That both China and Taiwan could converge on the final, "China, Nanjing" and "China, Taipei" mutual agreement so quickly (essentially within just a year: early 1979 to early 1980), after 20 years of living in separate worlds, seems a "miracle." And surprisingly, in the existing archives, the evolution of the geopolitical context is barely mentioned (except at the time of Taiwan's admission, for the reasons discussed at length in this chapter). Yet the whole story about Taiwan and China within the IAU is clearly *de facto* dominated by the history of USA–China relations, from the start of the problem to its solution, and explains why it lasted for so long and why it ended so quickly.

These relations can be summarized very briefly: In 1958, China and the USA were almost at war with each other, China and the USSR being allies. At that time, Taiwan was supported by the USA as a continuation of the "Republic of China" created by Sun Yat-sen in 1912 and now led by Chiang Kai-shek, who fled to Taiwan in 1949 after having been defeated by Mao Zedong in a bloody civil war. The relations between China and the USSR cooled down in the late 1950s for rather policy direction and political reasons; Mao's struggle for power gave rise to the Cultural revolution beginning in 1966, which essentially isolated China from the outside world. But progressively China gained support yes, ideological prestige is debatable. Even within the Socialist Bloc, most countries including in Asia, with the example of Vietnam still followed Moscow, so much so as to garner enough votes at the UN and become a member in September 1971, occupying the seat of Taiwan both at the Assembly and at the Security Council. As a result, Taiwan had to abandon the claim of representing all of China, while China stuck to its "One China" principle. With Nixon's visit to China in early 1972, and the subsequent recognition of China by an increasingly large number of nations, tensions between the USA and China eased progressively. Contacts between Chinese and Western astronomers resumed, but outside of the IAU. The issue of expelling Taiwan from the Unions was raised by UNESCO in 1973, but most of the Unions, including the IAU, opposed UNESCO. There was no significant evolution for some time, until Chiang Kai-shek and Mao Zedong died, almost at the same time (Chiang in 1975 and Mao in 1976). In Taiwan Chiang was succeeded at the head of the Guomindang party by his son, Chiang Ching-kuo, and some democratic reforms progressively appeared. In China, the struggle for power was complex, but at the IXth Congress of the Communist Party in 1977 Deng Xiaoping was eventually rehabilitated and gained support.

Then, the events accelerated: In a decisive step, the USA and China resumed diplomatic relations on January 1, 1979, and at the end of January, Deng visited the USA. On the other hand, the diplomatic relations between Taiwan and the USA took the form of just a treaty, the "Taiwan Relations Act" (which was in force until 2018). Deng's policy was to put emphasis on rather rapprochment with the USA and the West and self development, and *relations between China and Taiwan became authorized.* There was also an acquiescence rapprochment under Chiang Ching-kuo between Taipei and Beijing. The IAU–China–Taiwan relations could warm up quickly at that time, leading to a near-agreement in Montreal only a few months later. Of course, it did take the skills and diplomatic talents of people like Blaauw, Wayman, and others, not to forget the lifelong efforts by Goldberg, and of their Chinese and Taiwanese counterparts, to succeed and reach a final settlement. But much the same "dual adherence" agreements were reached at about the same time throughout the other Unions.

In summary, in spite of repeated pleas by ICSU, the IAU, etc., to "stay out of politics," the geopolitics of the China–USA relations were in reality "boundary conditions" severely limiting the possibilities of diplomatic action. *Without Deng's diplomatic opening to the world, the reconciliation between the IAU and China would have been impossible.*

8.5   Concluding Remarks

Chronologically, for the IAU the "China conundrum" was the first challenge of how to deal (or not to deal) with "divided countries." As mentioned, two more cases arose very soon after China's withdrawal: the split between the two Germanies in 1961, as a result of the Berlin Wall crisis, and the discreet admission of North Korea at the Berkeley General Assembly that same year. But these two cases turned out to be solved very easily by the IAU, because there was no dispute from one side to represent the other as one nation, and the UN could not interfere simply because ... none of these "divided countries" were members at the time! Thus, West and East Germany, after mutual concertation, adhered separately to the IAU at the Hamburg General Assembly in 1964 (whereas they were jointly admitted to the UN only in 1973), to resume a single IAU membership after their reunification in 1991, and South Korea adhered to the IAU, with no objection from North Korea, at the Sydney General Assembly in 1973 (whereas North and South Korea were jointly admitted to the UN much later, in 1991).

Similar to Korea, albeit much more laboriously, China (Nanjing) and China (Taipei) became officially "sister members" of the IAU in 1982. But by a twist of history, in the IAU listing of "National members," China (Nanjing) is said to have adhered in 1935 (the year of adhesion of the original Chinese Astronomical Society, based in that city), and China (Taipei) in 1959 (the year of its admission by the Executive Committee at Herstmonceux). In contrast, as if to erase an unpleasant episode, ICSU chose the same year for both national members, i.e., the original year of adhesion by the Academia Sinica: 1937. A twist here also: The adhering body for China is the China Association for Science and Technology (CAST), which was actually created in 1958, not in 1937 ...

Chapter 9
Epilogue

Right after the Montreal General Assembly, the Canadian astronomer Sydney van den Bergh, one of the new IAU Vice-Presidents, visited China on October 7–24, 1979, at the invitation of Peking's Academia Sinica. Contrary to his predecessors, who were in general rather kind and understanding in their appreciation of Chinese astronomy, van den Bergh produced a 25-page "trip report," found in the IAU archives (dated December 9, 1979), expressing much less candid views, and which apparently was never published. Its "Summary of conclusions" is perhaps worth quoting in full:

> Thirty years of almost complete isolation from the outside world and ten years of "Cultural Revolution" have had a devastating effect on astronomy in China. Astronomers of age under 35 (who are usually the most dynamic and innovative constituent of the scientific community) are virtually non-existent. The older astronomers, who will have to teach the new generation of graduate students, are with one or two exceptions almost totally out of touch with modern astrophysics. Only by training advanced students abroad can China hope to get back into the mainstream of astronomical research. It seems safe to predict that China will not become a major factor in world astronomy until well into the twenty-first century.

The diagnostic was terribly accurate, yet van den Bergh was too pessimistic: China has become for sure an important partner in astronomy, on the ground but also in space, with, e.g., the organization of the XVIIIth IAU General Assembly in Beijing (2012), which opens this book, and landing a rover for the first time on the far side of the Moon in 2021. Its astronomers are now often part of major international collaborations: On September 21, 2005, 25 years after China rejoined the IAU family, a ceremony was held in Mitaka, in the outskirts of Tokyo, where the National Astronomical Observatory of Japan (NAOJ) is located. The host was Norio Kaifu, Director General of NAOJ and future President of the IAU during my term as General Secretary (2012–2015).[1] Norio had played a key role in establishing the "*East Asian Observatory*" (EAO), an astronomy consortium of East Asian institutes

---

[1] Norio Kaifu (1943–2019). See his obituary in the IAU website: https://www.iau.org/news/announcements/detail/ann19023/?lang

T. Montmerle, Y. Zhou, *China and the International Astronomical Union*, Historical & Cultural Astronomy, https://doi.org/10.1007/978-3-031-01787-2_9

Fig. 9.1 Signing ceremony for the East Asian Core Observatories Association in 2005. From left to right: Zhao Gang (National Astronomical Observatories of China), Norio Kaifu (National Astronomical Observatory of Japan), Seok Jae Park (Korea Astronomy and Space Science Institute), and Paul Ho (Academia Sinica Institute of Astronomy and Astrophysics, Taipei). (Courtesy Kaz Sekiguchi, NAOJ)

built on the model of the European Southern Observatory (ESO), which operates a unique complex of high-altitude observatories in the Chilean Andes. The idea, as a starting point, was to set up a similar international collaboration to operate existing telescopes that were being replaced by larger ones on the Maunakea (extinct) volcano on the Big Island of Hawai'i.

The Mitaka ceremony was celebrating the signature of an agreement of collaboration (technically said a "Memorandum of Understanding") between a "core" of four founding institutes, called the *"East Asian Core Observatories Association."* The participants were the Directors or Presidents of astronomical institutes or associations of China (mainland and Taiwan), Japan, and South Korea (Fig. 9.1).

Today, the EAO has attracted more partners: Institutes from Vietnam, Thailand, Malaya, and Indonesia have joined in recent years. And some in India are considering also joining the consortium.

A masterpiece of "astronomy diplomacy," putting an end (hopefully) to over a century of hostility, war and destruction between the founding countries.

Appendix A: A "Personal Account" by D. Sadler (1976) of the Admission of Taiwan to the IAU

Beginning and End of the Handwritten *"Personal Account"* of Donald Sadler, 1976 (Internal Document Kept in the IAU Archives, with Permission)

Followed by Full Typescript by T. Montmerle

© The Author(s), under exclusive license to Springer Nature Switzerland AG 2022
T. Montmerle, Y. Zhou, *China and the International Astronomical Union*, Historical & Cultural Astronomy, https://doi.org/10.1007/978-3-031-01787-2

I very much hope that this matter can be settled by common sense on the part of both UNESCO and China; because I do not think that the IAU should expel Taiwan, unless it fails to pay its contributions.

C. Sadler

12 December 1976

The Admission of Taiwan to Membership of the International Astronomical Union[1]

A personal account by D. H. Sadler[2]

Explanatory Note. This account is written at the request of Professor Patrick Wayman, Assistant General Secretary of the IAU.[3] It is written without access to the minutes of the Executive Committee or to the Union's files of correspondence; and it is colored by personal bias and imperfect memory. It is unsuitable circulation outside the inner circles of the Union.

1. *Before the tenth General Assembly, Moscow 1958.[4]* Although I was acting as Assistant to the General Secretary (P. Th. Oosterhoff) during 1958, I was not a member of the Executive Committee and did not see all the papers. However, an application was received from the Academia Sinica in Taipei, Taiwan (calling itself the Republic of China) for membership of the Union. The President (A. Danjon) and the G.S. agreed that G.S. should request further information about the level of astronomical activity, even though this was not strictly necessary for consideration of the application according to the (then) Statute No.3. No reply had been received by the beginning of G.A. X, and the question was deferred by the Executive Committee to a later meeting.

---

[1] Typed transcript (by TM) of the original, handwritten document found in the IAU Archives (folder "Taiwan"). In case of doubt, a cautionary [?] appears. Some words were not quite legible; they are indicated by (xxx?). All footnotes are by TM.

[2] 1908–1987. Assistant General Secretary 1955–1958, General Secretary 1958–1961 and 1961–1964; Advisor 1964–1967; member of the Resolutions Committee 1967–1970 and 1970–1973 (See Sect. 3.4).

[3] 1976–1979.

[4] Here referred to as "G.A. X" (see, e.g., Fig. 3.4).

2. *During G.A. X.* The President convened, at very short notice, a meeting of certain members of the present and future Executive Committees, excluding nationals of U.S.S.R. and other "eastern" countries. The meeting was held in a private room at the Hotel Ukrainia and (in spite of a note from myself requesting restraint—I had been briefed by the Foreign Office !) was extremely frank. The President announced that he had received, by hand, a communication from a high authority in USA [I never saw this and do not know whether it was actually addressed from the State Department, or was passed by the Academy of Sciences] to the effect that the U.S. government would withdraw all support for the promised [?] invitation [for the Union to hold its G.A. XI in California in 1961] if the IAU did not admit the Republic of China to membership during G.A. X in Moscow. Leo Goldberg (nominated for the future Executive Committee), who I think may have known about this communication, made it clear that, without government support (especially as regards visas) it would be impossible to hold G.A. XI in USA. During the meeting, the opinion grew that the original application from Taiwan must have been originated by the U.S. State Department. There could be no doubt that the Union was being blackmailed in the most blatant manner. Danjon, quite justifiably, was furious and would, I think, have been prepared to reject the application outright.

However, it was agreed that

(i) Consideration of Taiwan's application should be deferred until the information requested had been received;
(ii) The communication from USA should be ignored;
(iii) Leo Goldberg should be authorized to "leak" this information to the appropriate authorities in USA (an unpleasant task !).

I do not know whether the U.S.S.R. members of the Executive Committee were aware of this particular piece of blackmail, though it could reasonably be assumed that the discussions in the private room at the Ukrainia Hotel could have been monitored. Certainly, no reference was made to it at subsequent meetings of the Executive Committee; but the General policy of the U.S. State Department was well known.

3. *At the nineteenth meeting of the Executive Committee.* This meeting was held at the Royal Greenwich Observatory, Herstmonceux in September 1959; it was attended by: J.H. Oort (President); L. Goldberg, O. Heckman, B.V. Kukarkin (accompanied by P.V. Verlikovsky as interpreter), B. Sternbek and R.H. Stoy (Vice-Presidents); D.H. Sadler (General Secretary), R.M. Petrie (Vice-President), and both advisors (A. Danjon and P. Th. Oosterhoff) were prevented by illness from attending.

Correspondence with the Academia Sinica and the National Committee of Astronomy in Taiwan had revealed a very small amount of astronomical activity (at about amateur level) and some real interest in the history of astronomy in Taiwan. Even this was more than that in at least one of the National Members (Venezuela) and clearly did not constitute an adequate reason for rejection of the application for membership. My interpretation of Statute No.3 was that there was no valid reason, or cause, for rejection, bearing in mind that:

(a) Taiwan was an adhering member of ICSU;
(b) It fulfilled all the technical conditions, including having a National Committee of Astronomy, recognized by the Academia Sinica;
(c) There was some independent astronomical activity;
(d) It could not be excluded, on the grounds that it was a "country," within the (xxx?) terms of the Statute, especially as no other "country" had de facto control of astronomical activity within Taiwan.

The question was discussed at considerable length, with the full knowledge that The People's Republic of China (under the name of China), which had been a National Member of the Union since 1935, would immediately withdraw if Taiwan was admitted; Oort, Goldberg, Stoy, and myself also knew that there was a high probability that the invitation to hold G.A. XI in Berkeley in 1961, already made and adopted [?], would be withdrawn if Taiwan's application was rejected. Most of the discussion centered on (d) above, but it was (a) that was unanswerable [?]: The IAU adhered to ICSU and its statutes, especially in respect to National Members, were linked with those of ICSU. It would have created much confusion if the IAU had interpreted "country" in a manner contrary to that which ICSU had already adopted.

Nevertheless, I felt strongly opposed to the blatant use of blackmail by the U.S. State Department, and tried to draft a resolution that might enable China to maintain its adherence to the Union. This resolution (printed in full in Trans. IAU, vol. XIB, page 33, 1962) was adopted unanimously by the Executive Committee after the main motion, to accept the application from Taiwan, was approved on a majority vote. The main motion was strongly opposed by Kukarkin, though I am sure that Verlikovski's verbal [?] translation of Kukarkin's Russian was both stronger and lengthier than the original.

4. *Before G.A. XI in Berkeley.* As foreseen, China formally resigned from the Union and, in spite of many efforts on the part of the President (Jan Oort) and myself, declined to participate in any activity of the Union or to accept any assistance (in the form of publications, technical information, etc.) from the Union. My personal relations with the Chinese astronomers had been excellent (they cooperated in the occultation program) and I had met them in Leningrad (at the opening of the Pulkovo Observatory in 1954).[5] But normal correspondence was impossible—all replies to letters, personal or otherwise, merely contained the re-iteration of the statement that Taiwan was an integral part of China. Whether publication of the Resolution adopted by the Executive Committee, would have helped is impossible to assess; I do not think it would, but I then considered it worth trying. I had planned to include it in the Information Bulletin No.21

---

[5][Note: also attended by Y.C. Chang (Zhang Yuzhe)]. See Zhou (2022).

(November 1959), but the President (taking the view that the policy had not been considered of approved by the General Assembly) requested that it be withdrawn at proof stage.

On the other side, Taiwan strongly objected to the Resolution and demanded that Taiwan should be known as The Republic of China, with short title China. We remained firm and eventually (after an interesting correspondence!) the position was accepted.

Nothing of relevance happened until early 1961 when two draft Resolutions, one by Czechoslovakia and the other by the U.S.S.R., were formally submitted for consideration of the General Assembly. Both (see Trans. IAU vol. XIB page 27) were essentially votes of censure on the Executive Committee, and demanded reversal of the decision to admit Taiwan.

Earlier Jan Oort and I had contemplated challenging the U.S. authorities by inviting, as guests, some Chinese astronomers to attend the General Assembly; (xxx?) with the U.S. sponsored ICSU resolution on "Political non-discrimination" host countries (for Union assemblies) gave an undertaking to admit all those qualified to attend by virtue of the Statutes of the Union (wording from memory). The President, according to IAU Statutes, could then invite guests. But, for fairly obvious reasons, we did not proceed with what might have been an interesting—but very dangerous for the IAU—test case.

5. *During G.A. XI.* The Executive Committee met before the General Assembly itself, and discussed (at great length—we finished well after mid-night) how the two Resolutions, by Czechoslovakia and U.S.S.R., should be presented to the General Assembly. All members and both advisors were present, but (if my memory is correct) the President-elect (also had been invited to attend) V.A. Ambartsumian, was not present. Eventually, it was agreed, due to the generous cooperation of Kukarkin (U.S.S.R.) and Steinberg[6] (Czechoslovakia), that:

(a) The two Resolutions should be combined;
(b) I should present the combined Resolution to the General Assembly stating, as objectively as possible, both sides of the case;
(c) The President should make an agreed statement of the reasons why the Executive Committee had decided to admit Taiwan; and (xxx?)
(d) This statement should include the previously unpublished resolution concerning nomenclature;
(e) The question should then be open for discussion before the Resolution was put to the ballot.

A full report of the meeting, of the President's declaration, of the resolution on nomenclature, and of the result of the voting is given in Trans. IAU Vol. XIB, pages 33–35.

---

[6] Sternberk.

All I can add to that report is to say that there was nearly a catastrophe. I had planned the General Assembly in detail and had given full notice of requirements to the Local Organizing Committee, regarding arrangement of the platform, positions of podium and microphones, seating plan (with names), requirement for a ballot box, etc. After the Inaugural Ceremony, having declined the invitation to lunch with Adlai Stevenson, I took all the papers required for the G.A. to Wheeler Hall in order to have everything prepared in good time; I was particularly aware that there would be large numbers of the press present, as the local newspapers had made news out of the (published) agenda. To my horror, I found that NOTHING had been prepared— there was merely a completely empty platform! All the senior members of the L.O.C. were at lunch with the Vice-Presidents, but I was lucky enough to find the chief (xxx?) superintendent of the University in his office. He said in effect "no problem," got teams of men on the spot within minutes and had the platform prepared—I believe he managed to get some flowers from somewhere!—in about 40 minutes, leaving 5 minutes to spare.

6. *At the ICSU meetings, October 1958.* Although not strictly relevant, the question of the two Chinas was, of course, brought up at the meeting of the Executive Board of ICSU at the Academy of Sciences, in Washington D.C. (almost in the shadow of the State Department) in October 1958. As far as I can remember, it was not discussed during the General Assembly.

Jan Oort and I represented the IAU. We were immediately appalled by the list of adhering countries, boldly announcing the membership of China (The Republic of China) with adhering organization the Academia Sinica, Taipei. We discovered that several other Unions (to which China adhered) had recently received applications from Taiwan, and I was informed that the (xxx?) letters of application had been drafted in the U.S. State Department by a scientist (who shall be nameless), who was actually a member of the ICSU Executive Board. I have never had any reason to doubt the accuracy of that information. During the meeting of the Board, Laclavère[7] (IUGG), Oort and I repeatedly attempted to raise the question of the representation of China, but were always faulted by the President of ICSU (L.V. Berkner)[8]; also, there was a majority (mainly from the ranks of the non-Union representatives) who clearly did not want the issue discussed in Washington. We were greatly [?] amused by the U.S. sponsorship of the ICSU declaration on "Political Non-Discrimination at Scientific Meetings"! (My proposal that the ICSU should ask the I.S.U. (International (xxx?) Union) to define the (xxx?) of China was not adopted!)

7. *General comments.* After nearly twenty years, I still feel that the action taken by the Executive Committee over Taiwan's application was the only correct one—in spite of the General revulsion against its political motivation. There is no doubt,

---

[7] Georges Laclavère, geodetist. Long time Treasurer of IUGG.

[8] American physicist and engineer. He was a member of the President's Science Advisory Committee (PSAC), which was created on November 21, 1957, by President of the United States Dwight D. Eisenhower, as a direct response to the Soviet launching of the *Sputnik-1* and *Sputnik-2* satellites.

however, that the Union's loss of China membership has been poorly repaid by the gain of Taiwan's membership.

I understand that pressure is being exerted on the IAU to expel Taiwan in order to permit China (The People's Republic of China) to resume its adherence. I do not know the background or the nature of the pressure, but I must call attention to the explanatory note to Statute 7 (see, for example, Information Bulletin No 23, page 26): "Once a country has become a member of the Union, there is no plausible ground for expelling it."

I do not entirely agree with this statement (due mainly to Jappel), but I do not think the Union should allow itself to be coerced into expelling a country for reasons *outside* its own concerns.

There are lots of problems. What happens when the U.K. devolution bill becomes law, or if Scotland becomes even more separated? Will the British government be responsible for astronomical activity in Scotland? The present absence of a definition of a "country" in its Statutes (as adopted at G.A. XIV, 1970) was a deliberate attempt to avoid the impossible task of specification that would involve political consider-ations. It may well be that, in changed circumstances, the IAU should align itself with ICSU and UNESCO and follow UN definitions, essentially identifying "coun-tries" as the National Members of the UN. This would mean following the UN lead in regard to South Africa, etc., but it would simplify the IAU problem: *Personally,* I would be opposed to so devolving decisions to the UN (or UNESCO), but the possibility of a unified policy between the scientific Unions under ICSU should not be excluded.

The position re. Taiwan should be clear: the IAU regards, and has always regarded, Taiwan as representing astronomical activity on the island of Taiwan and nowhere else. In so recognizing Taiwan's adherence, the IAU has not implied any authority to the government of the self-styled Republic of China other than that of the responsibility of the Academia Sinica in Taipei for astronomy in Taiwan. If a question should arise of some other organization (either inside or outside Taiwan) claiming that it is de facto responsible for astronomy within Taiwan, the IAU would have to decide which, if either, it should recognize. But I do not think that UNESCO, or the government of The People's Republic of China will deny the de facto responsibility of the Academia Sinica in Taipei; that China claims sovereignty over Taiwan is irrelevant to the IAU's relationship with the astronomers in Taiwan. If the IAU expelled Taiwan, the astronomers would not be represented. In this connection, it is worthy of note that there have been, to my knowledge, two cases in which two rival organizations *in the same country* claimed the right to adhere to the IAU-and, in one case, two rival representatives appeared at the General Assembly!

I very much hope that this matter can be settled by common sense on the part of both UNESCO and China; because I do not think that the IAU should expel Taiwan, unless it fails to pay its contributions.

(signed) D. H. Sadler

12 December 1976

Appendix B: IAU Archival Documents (1958–1979)

Appendix B1: Exchange of Letters between Y.C. Chang and J. Oort (1959)

Exchange of letters between Y.C. Chang and J. Oort *(facsimile)* after the Executive Committee decision to admit Taiwan to the IAU, 1959 (@ IAU, with permission)

© The Author(s), under exclusive license to Springer Nature Switzerland AG 2022
T. Montmerle, Y. Zhou, *China and the International Astronomical Union*, Historical
& Cultural Astronomy, https://doi.org/10.1007/978-3-031-01787-2

COPY

The Astronomical Society of
the People's Republic of China
Er Li Gou, Peking, China.

November 20, 1959

Professor J. H. Oort
President
International Astronomical Union
c/o Professor D. H. Sadler
Royal Greenwich Observatory
Herstmonceux Castle
Hailsham, Sussex, England

Dear Professor Oort,

We learn that the Executive Council Meeting of the International Astronomical Union held in England on September 7, 1959
has adopted the decision of accepting the so-called "Chinese
Astronomical Society" of the Chiang Kai-shek clique for membership
in the Union. We are greatly surprised by this decision.

As all the people of the world know, Taiwan is an inseparable part of Chinese territory, it is a province of China.
It is because of the armed occupation by the American imperialists
that Taiwan has not been yet liberated up to the present. The
Astronomical Society of the People's Republic of China is the only
legal organization to represent China in joining the IAU. The
Acceptance of the so-called "Chinese Astronomical Society" of the
Chiang Kai-shek clique by the Executive Council Meeting of the IAU is
evidently in keeping with the hoax of the American imperialists
of creating "two Chinas". It is illegal and wrong. Not only is it
a trespass on the legal rights ofChina in an international organization but also an obstruction of the normal development of international scientific cooperation.

Therefore, Mr. President, I ask you to clarify the above-
mentioned report. Should the report be authentic, I, on behalf of
the Astronomical Society of the People's Republic of China, hereby lodge our strong protest with you and insist that the Executive
Council of IAU rescind the illegal decision about the acceptance
of the so-called "Chinese Astronomical Society" of the Chiang Kai-
shek clique. Otherwise, the Astronomical Society of the People's
Republic of China will resolutely and definitely withdraw from
the IAU.

I await your reply.

Yours faithfully,

Y.C. Chang
President of the Astronomical
Society of the People's
Republic of China

C O P Y

Leiden Observatory
2 December 1959

Prof. Dr. Yu-Che Chang
Purple Mountain Observatory
Academia Sinica
Naking
China.

Dear Professor Chang,

In reply to your letter of 20 November 1959, and in reply to your
request for clarification, I formally acquaint you with the following decisions of the Executive Committee of this Union:

1) The application for membership of the Union from the Republic
   of China (Taipei, Taiwan) be accepted for adherence under Statutes, Section II, No. 3(b), with the Astronomical Society of
   the Republic of China (Taiwan) acting as the National Committee
   of Astronomy, appointed by Academia Sinica (Taiwan) and with
   membership in Category 1.

2) Whereas there is the possibility of confusion in the descriptions
   variously accorded to the two Republics of China, both of which
   adhere to the Union, and

   in order to indicate unambiguously the two geographical areas in
   which independent astronomical activity has been developed,

   it is resolved that short names, as indicated below, shall be
   used for all purposes of the Union, with the exception of formal
   correspondence and descriptions in which the use of the official
   titles of the two countries is desirable,

   1. Official title of the country: The People's Republic of China
      Site of Government: Peking, China
      Adhering Organization: National Committee of Astronomy
                                      (Nanking)
      Date of Adherence: 1935
      Geographical area in which
      Astronomers are represented: The mainland of China
      Short name: China (in French: Chine)

   2. Official title of the country: The Republic of China
      Site of Government: Taipei, Taiwan
      Adhering Organization: The Astronomical Society of the
                                      Republic of China (Taipei), recog-
                                      nized as a National Committee of
                                      Astronomy by the Academia Sinica,
                                      (Taipei), through which organiza-
                                      tion the Republic of China adheres
                                      to the International Council of
                                      Scientific Union
      Date of Adherence: 1959
      Geographical area in which
      Astronomers are represented: Taiwan (Formosa)
      Short name: Taiwan (in French: Taiwan)

The decisions of the Executive Committee cannot be rescinded. In this case the Astronomical Society of the Republic of China (Taipei) has already been informed of these decisions.

With regard to the views expressed in your letter I think it is desirable to affirm that the only considerations underlying the decision to admit Taiwan to membership of the Union were scientific, namely the desire to fulfil the objects of the Union by the encouragement of Astronomy in a territory which is, as far as the Executive Committee is aware, not otherwise directly represented in the Union. The executive Committee understands that the astronomers in Taiwan are not represented by the Astronomical Society of the People's Republic of China, of which you have the honour to be President; they will therefore, in practice, have no means of being represented in the Union or becoming members of the Union. The acceptance of membership of Taiwan will enable astronomers in Taiwan to participate in the international organization of astronomy and will it is hoped encourage the development of Astronomy in Taiwan.

I must emphasize that the representation covered by the Astronomical Society of the Republic of China in Taipei is limited to those astronomical activities in Taiwan that are at present (and for whatever reason) independent of the guidance, supervision and representation of the Astronomical Society of the People's Republic of China in Peking. I am sure that you as President of one of the foremost Astronomical Societies in the world would not wish to deny the benefits of membership of the I.A.U. to any group of astronomers.

The application from the Astronomical Society of the Republic of China in Taipei formally conformed to the Statutes of the Union and to the Statutes and Directives of the International Council of Scientific Unions, to which the I.A.U. adheres.

According to our statutes it would, in fact, have been impossible for us to refuse Taiwan's application after it had been indicated that some independent astronomical activity existed in this territory, as you will see from the following sentences of article 3 of the Statutes:

"Un pays qui adhère au Conseil International des Unions Scientifiques peut adhérer à l'Union (a) par l'organisme par lequel il adhère au Conseil International, ou (b) par un Comité national d'astronomie formé ou autorisé par cet organisme.
Dans le mot "Pays" sont compris les Dominions, les protectorats Diplomatiques, ainsi que les territoires ayant une activité astronomique indépendante."

Certainly, the astronomical activity in this territory appears at present to be very limited. But this is no reason for refusing admission. As you may remember, several countries have in the past been admitted, in which astronomical research was only in its very first beginning. Former Executive Committees have felt, as the majority of the present Committee did, that membership of the I.A.U. might contribute to facilitate developing astronomical research. At any rate, the Executive Committee felt they could not discriminate.

I am certain that I can speak for all members of the Union when I
say that the withdrawal of the People's Republic of China from membership of the Union would be a severe loss to Astronomy and a bad blow
to international cooperation in the field of science which has so far
led the world. I ask you to reconsider carefully your views in the
light of this letter and of my assurance that the sole and only purpose for the admission of Taiwan is to help the astronomers working
in Taiwan and that astronomers throughout the world are anxious to
continue that co-operation with astronomers in the People's Republic
of China, for whom we all have high admiration and regard.

Yours sincerely,

signed: J. H. Oort.

Appendix B2: Radioteletype Transcript Announcing the Withdrawal of China from the IAU (1960)

"Radioteletype" transcript of a radio broadcast by the Xinhua News Agency announcing the withdrawal of China from the IAU, February 6, 1960 (@ IAU, with permission)

CPR WITHDRAWS FROM ASTRONOMICAL UNION   Feb. 9, 1960

Peking, NCNA, Radioteletype in English, to Europe and Asia, Feb. 6, 1960
0800 GMT-W

(Text) Peking, Feb. 6--The Astronomical Society of the People's Republic of China strongly protests the decision of the International Astronomical Union (IAU) to accept the so-called "Astronomical Society of the Republic of China" of the Chiang-Kai-shek clique as its member, in line with the U. S. imperialist intrigue of creating "two China's". The protest was made in a letter sent yesterday to J. Oort, chairman of the executive committee of the IAU, by Chang Yu-che, president of the Astronomical Society of the People's Republic of China. In the letter, the Astronomical Society of the People's Republic of China announced its immediate withdrawal from the IAU and all its affiliated commissions, and a discontinuance of all connection with them.

The organization of Chinese Astronomical workers joined the IAU as early as 1935. Since the founding of the Chinese People's Republic, the Astronomical Society of the People's Republic of China has continued to cooperate with this international organization and to enjoy its membership status. It took part in the ninth IAU congress held in Dublin in 1955 and the 10th congress held in Moscow in 1958. But the session of the IAU Executive Committee held in London in September 1959, ignoring the fact that the Astronomical Society of the People's Republic of China is the sole lawful organization of China astronomical workers, and despite the firm opposition raised by the representatives of the Soviet Union and other countries, illegally accepted the "Astronomical Society of the Republic of China" of the Chiang Kai-shek clique as its member.

Upon receipt of this news, the Astronomical Society of the People's Republic of China, in a letter last November, lodged a strong protest with Professor Oort and demanded that the IAU Executive Committee rescind this illegal decision. This just demand was unreasonably rejected by Oort in his reply last December. In view of this, the Astronomical Society of the People's Republic of China has decided to withdraw from the IAU.

In his letter to Oort on Feb. 5, Chang Yu-che said that Oort in his letter "openly put on a par the great People's Republic of China and the Chiang Kai-shek clique which has long since been disowned by the Chinese people, and with ulterior motives quoted article three of the statutes of the union as "grounds" for representing Taiwan as a country. Evidently, this is in line with the U. S. imperialist intrigue of creating "two China's".

"Against this, I feel compelled once more to lodge our strong protest with you. I have repeatedly explained to you: There is only one China in the world; that is, the People's Republic of China. Taiwan is an inalienable part of Chinese territory since ancient times. And the present situation is only the result of the armed occupation by American imperialists. The American imperialists have, in recent years, been actively engaged in the plot of creating "two Chinas" with a view toward legalizing their forcible occupation of China's Taiwan. The Taiwan question is China's internal affair and the Chinese people are determined to liberate Taiwan. We are firmly against the conspiracy of the American imperialists to create "two China's" and we will oppose it to the end".

Feb. 9, 1960

The letter pointed out: "The Astronomical Society of the People's Republic of China is the sole lawful organization of the Chinese astronomical workers. It represents the astronomical workers in Taiwan just as it does the astronomical workers in all provinces of the country. It is only due to the obstruction by the American imperialists that we are for the time being not in a position to keep contact with the astronomical workers in Taiwan. It is (an open?) inter-ference in China's internal affairs on the part of the IAU to take this as a pretext to deny the right of our astronomical society to represent the astronomical workers in Taiwan. This is utterly intolerable to us."

The letter concluded: "The admittance of Taiwan to membership is a thing which goes far beyond the scope of "science", and reduces the IAU to a mere tool in the political intrigue of "two China's". This is not only an act of hostility to the Chinese people, but also runs counter to the will of fair-minded scientists in the IAU itself. In view of the fact that the IAU, by ignoring our stern protest and persisting in a wrong decision, has destroyed the cooperative relations between itself and China, I, on behalf of the Astronomical Society of the People's Republic of China, hereby declare our immediate withdrawal from the IAU and all its affiliated commissions, and a discontinuance of all connections with them".

Appendix B3: Statement of the People's Republic of China after the ICSU General Assembly (1974)

Statement of the Chinese Scientific and Technical Association of the PRC after the General Assembly of ICSU, Istanbul, September 1974 (@ IAU, with permission)

中 华 人 民 共 和 国 科 学 技 术 协 会

THE SCIENTIFIC AND TECHNICAL ASSOCIATION
OF THE PEOPLE'S REPUBLIC OF CHINA
PEKING, CHINA

T in 1642

STATEMENT OF THE SCIENTIFIC AND TECHNICAL ASSOCIATION OF THE PEOPLE'S REPUBLIC OF CHINA

November 1, 1974

Recently the International Council of Scientific Unions held its 15th General Assembly in Istanbul, at which its relationship with the People's Republic of China was discussed. During the discussions the representatives of many countries upheld justice and strongly demanded the restoration to China of her lawful rights in ICSU. For this we wish to express to them our heartfelt thanks. But this Assembly, under the manipulation of a few leading officials, adopted on September 25, 1974 a resolution providing that "an appropriate institution in the People's Republic of China be again invited to become a member of ICSU" while preserving the membership of the Chiang Kai-shek clique. This resolution diametrically runs counter to the true desire of the vast scientific workers of ICSU. The Scientific and Technical Association of the People's Republic of China and all Chinese scientific workers express their firm opposition to this resolution.

As is well known, there is only one China in the world, that is, the People's Republic of China. Taiwan is one of the provinces of the People's Republic of China. The Chiang Kai-shek clique, which is entrenched in Taiwan, has long been repudiated by the Chinese people and any institution under its control has no qualification to represent China to participate under whatever name in ICSU or any other international organization. The Scientific and Technical Association of the People's Republic of China is the sole lawful organization representing all the scientific workers of China. Any plot to create socalled "two Chinas" or "one China, one Taiwan" can never be tolerated by the Chinese people.

It must be pointed out that the Unesco Executive Board adopted, on October 8, 1973, at its 93rd session a decision which "urges international non-governmental organizations which maintain relations with Unesco and in which bodies or elements linked with Chiang Kai-shek participate, having illegally usurped the name of China, to take measures to exclude them immediately and to break off all relations with them." And ICSU statutes stipulate: "A National Member should not normally be considered as ready for full membership, unless it... has a national body representing the country's scientific activities...". Accordingly, ICSU ought to have immediately expelled elements of the Chiang Kai-shek clique and restored the lawful rights of the Scientific and Technical Association of the People's Republic of China in it. Nevertheless, certain leading officials of ICSU disregard the decision of the Unesco Executive Board and the provision of the ICSU statutes and, while expressing "an appropriate institution in the People's Republic of China be again invited to become a member of ICSU", insist on illegally retaining elements of the Chiang Kai-shek clique so that they can hang on and continue their activities in ICSU. This is, in reality, creating "two Chinas" or "one China, one Taiwan" and constitutes an interference in China's internal affairs. Those leading officials of ICSU try to justify their action by pleading "universality" and "no involement in politics". This is untenable. The fact is that their absurd decision precisely runs counter to the principles they advertise.

The Scientific and Technical Association of the People's Republic of China solemnly states that those leading officials of ICSU must discard their wrong position, immediately take measures to put in practice the correct decision of the Unesco Executive Board, expel elements of the Chiang Kai-shek clique and break off all relations with them, and restore to the Scientific and Technical Association of the People's Republic of China all its lawful rights in ICSU.

At present, to develop friendly exchanges with Chinese people and scientific workers has become a strong desire of the people and scientific workers of all countries and an irresistible trend. We are confident that, through the joint efforts of the scientific workers of the whole world, the just position for restoring the lawful rights of the People's Republic of China in all international scientific organizations will eventually win out all obstacles.

中华人民共和国科学技术协会声明

（一九七四年十一月一日）

不久前，国际科学协会理事会在伊斯坦布尔召开了第十五届全体大会。 这次会议讨论了国际科协与中华人民共和国的关系问题。在讨论中，很多国家的代表主持正义，强烈要求恢复我国在国际科协中的合法权利。对此，我们表示衷心的感谢。但是这届大会在极少数领导人的操纵下，于今年九月二十五日竟然通过在保留蒋介石集团会籍的情况下"再次邀请中华人民共和国的适当机构参加国际科协"的决议。这是完全违背国际科协广大科学工作者的真正意愿的。对此，中华人民共和国科学技术协会和全中国科学工作者表示坚决反对。

众所周知，世界上只有一个中国，就是中华人民共和国。台湾是中华人民共和国的一个省。盘踞在台湾的蒋介石集团早已被中国人民所唾弃。 它的御用机构根本没有资格代表中国，也没有资格以任何名义参加国际科协或任何其它国际组织。中华人民共和国科学技术协会才是代表全中国科学工作者的唯一合法组织。任何制造所谓"两个中国"或"一中一台"的阴谋，都是中国人民绝对不能容忍的。

必须指出，联合国教科文组织执行局第九十三届会议一九七三年十月八日通过的决议，"敦促所有非法盗用中国名义的、同蒋介石集团有联系的团体或分子参加的、同教科文组织有联系的非政府性国际组织采取措施，立即开除他们的会籍，停止与他们的一切联系。"

国际科协会章也规定:"一个国家会员必须……有一个代表该国科学活动的国家机构……才能被认为具有正式会员资格。"据此,国际科协理应立即驱逐蒋介石集团分子,恢复中华人民共和国科学技术协会在国际科协中的合法权利。国际科协某些领导人却无视联合国教科文组织执行局的决议和国际科协会章的规定,一面"再次邀请中华人民共和国的适当机构参加国际科协",一面继续非法容纳蒋介石集团分子,让其赖在该组织中继续活动。这种做法实际上是在制造"两个中国"或"一中一台",是对中国内政的干涉。国际科协的某些领导人以"普遍性"和"不介入政治"来为他们上述行为进行辩解,这是站不住脚的。事实正相反,他们采取的荒谬决定同他们所标榜的"原则"是完全背道而驰的。

中华人民共和国科学技术协会郑重声明,国际科协某些领导人必须改变他们的错误立场,立即采取措施实现联合国教科文组织执行局的正确决议,驱逐蒋介石集团分子,并断绝与他们的一切来往,从而恢复中华人民共和国科学技术协会在国际科协中的一切合法权利。

当前,发展同中国人民和科学工作者的友好交往已成为各国人民和科学工作者的强烈愿望,这是谁也阻挡不了的巨大潮流。我们相信,在全世界广大科学工作者的共同努力下,恢复中华人民共和国在所有国际科学组织中合法权利的正义主张终将克服一切阻挠而获得胜利。

Appendix B4: The "Montreal Agreement" (1979)

The "Montreal Agreement" (August 22, 1979) endorsing the admission of China to the IAU: Exchange of letters between A. Blaauw and Y.C. Chang (Zhang Yuzhe) *(originals),* and Draft resolution to be submitted to the General Assembly (@ IAU, with permission)

Montreal, August 22, 1979

To Prof. A. BLAAUW, President
of the International Astronomical Union
c/o General Assembly of IAU 1979
Montreal, Quebec, Canada

Dear Professor Blaauw

On behalf of the Chinese Astronomical Society, whose
delegates have convened with IAU Officers during the 17th General
Assembly of IAU at Montreal, I herewith confirm that,

considering that both the Chinese Astronomical Society and
the IAU believe that the question of China's representation in the
IAU needs urgent attention;

noting with satisfaction that the IAU wishes to find a solution
to the question on the principle of one China, embracing the
Mainland and Taiwan;

understanding the nature of complexity of such an endeavor in
the present circumstances;

considering also the desire expressed by the IAU that the
Chinese Astronomical Society is welcome to return without preventing
the astronomical organization as well as astronomers in Taiwan from
participating in the IAU;

reiterating that the Chinese Astronomical Society opposes any
solutions which can suggest the existence of "two Chinas" or "one
China, one Taiwan";

the Chinese Astronomical Society proposes to the IAU
Executive Committee

a) that for the time being dual adherence should be introduced,
   the adherence and the adhering bodies to be specified as:

China ⟨
      Chinese Astronomical Society (Purple Mountain Observatory, Nanking)

      (a name for the adhering organization representing
      the astronomical community of Taiwan; for this name
      we propose "Chinese Taiwan Astronomical Society" or,
      provisionally, as a reference "Astronomy Taiwan, China"

- 2 -

the name to be established by the IAU Executive
Committee in consultation with the two adhering
organizations);

b) that for the time being the financial contribution of China
   be composed of one component due from the Chinese Astronomical Society (Nanking) and one component due from (the name
   to be defined as under (a));

c) that for the time being each of these adhering bodies nominates a representative authorized to vote in the General
   Assembly. In accordance with article VII, 11 of the IAU Statutes, on questions involving the budget of the Union, the number of votes of these representatives will be one greater than
   the number of its category as defined in article 8, and on
   questions of administration, not involving the budget, each
   representative will have one vote.

Yours sincerely,

张 钰 哲 Y. C. Chang

Y.C. Chang, President
Chinese Astronomical Society

August 22, 1979

To Prof. Y.C. CHANG, President
of the Chinese Astronomical Society
c/o IAU General Assembly 1979
Montreal, Canada

Dear Professor Chang,

In reply to your letter of August 22, 1979 I have the pleasure to inform you as follows:

The Executive Committee has taken note with great satisfaction of the proposal contained in your letter, recognizing the Chinese Astronomical Society's view leading to this proposal. It wishes to express its high appreciation for the constructive spirit in which the delegation of your Society has contributed towards restoring full participation of China in the IAU.

The Executive Committee therefore will be glad to take up the problem of deciding on the name to be used for the adhering body on behalf of the astronomical community of Taiwan. It will do so in the near future in consultation with the astronomical community of Taiwan and with your Society, and possibly benefiting also from the experience gained meanwhile by other international scientific unions working similarly towards the full participation of China.

The Executive Committee, in view of the discussions held during this Assembly, hopes and expects that this course will soon lead to the formal restoration of full Chinese participation and would be extremely pleased to see the valuable participation of members of your Society in the scientific discussions at this Assembly, continued at the forthcoming scientific meetings of the IAU. To this purpose, the General Secretariat will supply your Society with all relevant documentation.

Yours respectfully,

A. Blaauw, President
International Astronomical Union

The General Assembly,

in view of the reported progress in restoring full
adherence of China;

noting the letters on this subject exchanged between
the President of the IAU and the President of the Chinese
Astronomical Society, Prof. Y.C. Chang;

authorizes the Executive Committee to implement the
anticipated conclusion of the current negotiations subject
to ratification by the General Assembly at its next meeting.

Appendix C: A Visit to China Astronomy in 1979

Pierre Léna
  Observatoire de Paris (Meudon) and Académie des sciences, Paris, France

1. The Context

December 1978: Deng Xiaoping became the leader of China and began his economic reforms. The Great Proletarian Cultural Revolution, with its destructions, was over. That same year, Vietnam had invaded Cambodia, provoking the end of the Khmer Rouge's regime, until then supported by China. In retaliation and for a month after February 1979, China had fought a short punitive war against Vietnam.

In 1961, astronomers from the Republic of China (i.e., Taiwan) were admitted as members of the International Astronomical Union (IAU), causing the withdrawal of the People's Republic of China (PRC, i.e., mainland China). A decade later, in 1971, the PRC was admitted into the United Nations, a decision which provoked the expulsion of Taiwan from the UN and from UNESCO, but which was neither followed by the IAU, nor by most other scientific unions that were members of ICSU (the International Council of Scientific Unions). In the new Chinese political context of 1978, the long process initiated by the IAU to accommodate both the PRC and Taiwan was beginning to show progress, but two more years would be needed before a satisfactory agreement could be reached at the IAU General Assembly in Montreal (1979), and ratified at the following General Assembly in Greece (1982) after a modification of the IAU Statutes, giving astronomers of these two parts of China their place within the IAU and the possibilities to rejoin the growing international exchanges and projects in astronomy.

Since the work of Matteo Ricci, beginning 1553, and his Jesuit companions in China, Chinese astronomy has fascinated the West. What would be its future after these destructions, in this new political and scientific context when in the West great observatories on the ground and in space, as well as new physics, were accumulating discoveries[9]? During the 1970s, eminent astronomers began to visit mainland China: In 1973, George Miley from Leiden[10]; in 1977, a US National Academy of Sciences delegation headed by Leo Goldberg[11]; the same year, Charles Fehrenbach from France made a visit, followed by a movie maker, Marceau Ginesy.[12] In early 1979, Francis Graham Smith, director of Greenwich Observatory, visited the observatories, and the IAU Assistant General Secretary Patrick Wayman met with Chinese officials[13] to prepare the rejoining of China at Montreal. In the spring of 1980, Edwin Salpeter and Geoffrey Burbidge lectured in Beijing.[14] Conversely, in October–November 1978, the astronomer Mrs. Ye Shuhua, born 1927, then Vice-President of the Chinese Astronomical Society and member of the 5th National Conference of the National People's Congress, chaired by Deng Xiaoping, visited France as head of a delegation of the Chinese Academy of Sciences (CAS, 中国科学院).[15,16] She would be part of the six-member Chinese delegation invited to attend the Montreal General Assembly in August 1979.

Ye Shuhua's work in astrometry had led in 1965 to give to China an excellent time standard, related to the Universal Time with an accuracy which ranked second in the world[17]: This was an exceptional international achievement at a moment. The reputed role of French astronomers and the Observatoire de Paris in astrometry and

---

[9]Interest in the West for contemporary Chinese science, certainly with some geopolitical concerns, was emerging earlier. Cf. *Directory of selected scientific institutions in Mainland China*, Survey and Research Corp. Hoover Institution Press, Stanford University, 469 pages, 1970.

[10]*Sky and Telescope*, March 1974.

[11]*Astronomy in China: a trip report of the American astronomy delegation.* National Academic Press, 1979.

[12]*Astronomie en Chine*, by Marceau Ginesy, *L'Astronomie* (magazine), 1978. The same material was used for a TV show on the channel TF1.

[13]Wayman P.A. 1979, *An astronomer's visit to China*, Irish Astronomical Journal, Vol. 14, pp. 66–68.

[14]In 1978, the French Embassy in Beijing had appointed a Commercial Counselor just married, whose wife was a young astrophysicist, Florence Durret, who was just graduated in astrophysics from our school in Paris. As she wished to begin a PhD during her stay in China, Ye Shuhua was decisive in getting Wang Shouguan, director of Beijing Observatory, to host her for a couple of years. In our correspondence, Florence Durret mentions the lectures of Salpeter and Burbidge, which she attended, and the studies of magneto-hydrodynamics phenomena which she could then begin jointly with Meudon on her PhD topic. Two years later, Ye Shuhua was decisive again in facilitating the visit to Chinese observatories of another of our graduate students, François Viallefond, who was engaged with a young Chinese woman he later married. Born in a small village near Zhongshan, she could friendly communicate in Cantonese with Ye Shuhua, whose birthplace was the nearby city of Shun De (F. Viallefond, personal communication).

[15]Sometimes still called "Academia Sinica" at that time (see Fig. C1), although strictly speaking the original "Academia Sinica" of the Republic of China (founded in 1928) moved to Taiwan in 1949, where it continues to bear its original name.

[16]Report by Pierre Charvin (1980), see below, Foonote 21.

[17]叶叔华: 被尊称为"先生"的传奇女性. *Sina (in Chinese). 2014/09/15. Quoted by* https://en. wikipedia.org/wiki/Ye_Shuhua.

time measurement made indeed her 1978 visit to Paris a quite natural one. Visiting Paris Observatory and its CERGA[18] astrometric station near Nice, she had met two eminent French astronomers, Jean Delhaye and Jean Kovalevsky, both being known as astrometry specialists, i.e., the methods to measure with utmost accuracy the positions of stars in the sky and to follow their tiny motions. Astrometry was indeed related to the diurnal rotation of the Earth, and the definition of time. It was also involved in *Very Long Baseline Interferometry* (VLBI) at radiofrequencies, a combination of distant radio telescopes which she was pioneering in China. It was during this visit to France that I met her for the first time, the beginning of friendly exchanges which would be pursued over the next decades.

2. Visit of a French Astronomical Delegation to China (1979)

In early 1979, at Ye Shuhua's initiative after her return from France, the CAS sent an invitation, proposing a visit of a French delegation (Fig. C1). It was handled in France by the Office for international relations within CNRS.[19] The visit was to last two weeks (October 21–November 4) and include stays in Kunming, Shanghai, Nanjing and Beijing.

Jean Delhaye and Jean Kovalevsky, both members of the French Académie des sciences, led the delegation, having completed it with Pierre Charvin and myself (Fig. C2). It is interesting to note that, at exactly the same moment, a seven-member delegation from the CAS, headed by Li Dezhong (李德仲), Director of its National Space Science Center, visited in Paris the Centre national d'études spatiales (CNES), to get information on the French space program.

Jean Delhaye[20] had been the Director of the Paris Observatory from 1967 to 1971, then had been placed in charge of a new body, INAG,[21] within the Centre National de la recherche scientifique (CNRS), to coordinate astronomical research in France, build a large optical telescope in Hawaii (in cooperation with Canada and University of Hawaii), and oversee the French participation to the European Southern Observatory (ESO) created in 1962. Jean Kovalevsky (1929–2018) had pioneered the early use of computers, in order to improve the accuracy in describing

---

[18] CERGA: *Centre d'études et de recherches géodynamiques et astronomiques.*

[19] The French national research agency.

[20] Jean Delhaye (1921–2001) was also the president of the Bureau des Longitudes, a respected institution created during the French Revolution by the *Convention Nationale* (25/06/1795), initially to follow the progress in the determination of longitudes, a critical problem indeed for the navigation at sea, and later involved in the measurement of positions in the Solar System. Delhaye's contributions to the motion of stars in the Galaxy played a fundamental role in the creation of an international data center in Strasbourg (*Centre de Données Stellaires*) and in the preparation of the first astrometric satellite *Hipparcos* (which would be launched in 1989). https://fr.wikipedia.org/wiki/Jean_Delhaye

[21] INAG: *Institut National d'Astronomie et de Géophysique.*

中 国 科 学 院
上 海 天 文 台 (徐家汇部分)
SHANGHAI OBSERVATORY,
(ZI-KA-WEI SECTION)
ACADEMIA SINICA,
SHANGHAI, CHINA.

June 12, 1979.

Dear Prof. Lena:

It is so sorry for the delay of your visit. I have learned that the Academy of Sciences of China has not yet got information of the date of your coming, and they think that it would be somewhat difficult to arrange the schedule for five observatories visiting in so short a time, because our transportation is not so convenient as in Europe, for example, there would be no daily flight from Peking or Shanghai to Kunming. They are trying their best to manage it, but it seems you could stay in Peking for only 1-2 days, I think it would be a pity. Would you mind to let me know where you particular wish to stay, I will try to discuss with them.

Regarding your lectures, I would suggest you to talk about the instrumentation and operation for far infrared observations with a balloon-borne telescope, positioning of the telescope, and the pointing as well as stabilization of it, the analysis and processing of the results etc.

Madame Durret is contacting with the Peking Observatory and the Peking University now. I wish her to hear the chance of staying there. With my best wishes.

Yeh Shu-Hua
in Peking

Fig. C1 A letter dated June 12, 1979, and preparing the visit, among the many, usually handwritten, exchanged between Ye Shuhua and Pierre Léna over the months and years to come. Note the letterhead of the stationery, "*Academia Sinica,*" instead of the official name "*Chinese Academy of Sciences*" (see Footnote 14), and the name "Zi-Ka-Wei" for the section of the Shanghai Observatory, which is now written *Xujiahui* (徐家汇) in the pinyin transcription. (Archives P. Léna, with permission of Ye Shuhua)

Fig. C2 The delegation of astronomers from France, during a visit to the Summer Palace in Beijing. *Left to right*: Jean Kovalevsky, the interpreter (中文翻译), Jean Delhaye, Pierre Léna, Pierre Charvin. (Photo: P. Léna)

the motions of natural satellites in the Solar system. He had just created the CERGA and installed there a laser sending light to mirrors placed on the Moon's surface by the *Apollo* missions, then receiving the echo and measuring with utmost accuracy the constantly changing Earth–Moon distance. He also was playing a key role in the preparation of the European *Hipparcos* satellite. Pierre Charvin,[22] a solar astronomer and deputy director of INAG and working with Jean Delhaye, was playing a key organizational role of astronomical research in France, and would soon become the Director of the Paris Observatory for ten years (1981–1991). Pierre Charvin and myself had flown together on board of the supersonic *Concorde* aircraft in 1973, organizing the flight and observing a solar eclipse, with a full totality lasting 74 minutes.[23] The youngest member of the team was myself (born 1937), Professor of astrophysics at the University Paris-7, engaged in infrared astronomy and chairing the Scientific Committee of the European Southern Observatory, in order to prepare the decisions for the next European *Very Large Telescope* (VLT) (see the delegation visiting Beijing, Fig. C2). Having built ties in Tucson (Arizona) with Leo Goldberg,

[22] "En souvenir de Pierre Charvin" *Bulletin Intérieur de l'Observatoire de Paris*, n°867bis, 21/6/ 1990. See also: "Hommage à Pierre Charvin." In *Lettre d'information interne de l'Observatoire de la Côte d'Azur*, n°4, Février 1990. Pierre Charvin was the Administrator then Director of Paris Observatory for ten years (1981–1989).

[23] P. Léna, *Racing the Moon's shadow with Concorde 001*, Springer, 2016. Translated from French, *Concorde 001 et l'ombre de la Lune*, 2014.

IAU President (1973–1976) and director of NOAO, I was aware of the difficulties related to the participation of mainland Chinese astronomers to the IAU.

On October 20, 1979, we left Paris for Beijing, with a refueling stop of the Air France plane at Karachi airport. Our visit in China lasted two full weeks, first flying to Kunming then to Shanghai, where a train took us to Nanjing, then a slow night train to Beijing, as all these places had astronomical observatories and represented the "standard tour" of Chinese astronomy. It meant for us an incredible discovery of a country reopening after a decade of seclusion from the West, plunging into a long history of astronomy and experiencing everywhere a strong will to re-enter into world science. An extensive report[24] to French authorities was produced early 1980. Here, I limit my comments to some aspects of the visits and more personal impressions, as I could present these a few months later in a seminar[25] given at Paris Observatory.

Everywhere, we received a warm reception from astronomers and engineers full of curiosity, many of them speaking a reasonably good English, the language into which many of their publications were available in bilingual versions. Despite the context of nearing to a final solution to the IAU crisis, our exchanges did not cover political issues, except if somewhat related to the current difficulties of astronomical research in the country, such as the loss of several generations of students. Our contacts with the uniformly dressed population itself were rather limited, given the tight program and the firm guidance of our interpreter. Nevertheless, we could discover the mountainous and beautiful countryside around Kunming Observatory (2020 m) in Yunnan, the peasants at work on the way to Zo Se—the observing station of Shanghai observatory—or when driving to the Xinglong observing station (960 m), 150 km north of Beijing. In Beijing, filled with an incredible traffic of bicycles, the Democracy Wall was in full activity, covered with dazibaos of all sizes and writings, with crowds reading and commenting on them (Fig. C3).

3. Some Observations during the Visit

In terms of telescopes, China was lagging behind European astronomy: A 1-m optical telescope, built by Zeiss in Iena (East Germany), had just been inaugurated in Nanjing; a similar one in Yunnan Observatory near Kunming; like the radio telescopes at Shanghai and Beijing, the solar instruments at Nanjing were standard products of the 1960s, and their scientific exploitation was rather limited, although a millimeter wave detector was under development in Nanjing. No modern sensitive

---

[24] P. Charvin, *Compte-rendu de mission en Chine*, 28 pages, no date (most likely beginning 1980). Annexes from Jean Kovalevsky and Pierre Léna. [Archives CNRS/INAG and personal archives P. Léna]. See also the short report *Rapport préliminaire sur la visite en Chine d'une délégation d'astronomes français*, J. Delhaye, 15/11/1979, 3 pages [archives INAG].

[25] P. Léna. "*Une visite des principaux centres astronomiques en Chine.*" Séminaire de l'Observatoire de Meudon, 25/2/1980.

Fig. C3 Late October 1979, crowds are reading the dazibaos displayed on the Democracy Wall in Beijing (Photo: P. Léna)

optical detectors,[26] practically no infrared detectors, no computers were available to pave the path to discoveries, as experienced in the West since one or two decades. A 1.56-m optical telescope in Shanghai (Zo Se) was in an early planning stage. Only in VLBI was China implementing a modest network of three radio telescopes. Giving seminars on the progress in France and Europe, we could perceive how aware our Chinese colleagues were of this growing gap, how well informed they were of the most recent developments outside China, and how willing they were to close the gap as soon as possible with the Chinese traditional tenacity. At a time where a number of new optical telescopes of 3.6-m to 4-m aperture were just becoming available to Western astronomers in both hemispheres, fully equipped with modern detectors, this gap was soon going to broaden. The construction of a 2- to 4-m class instrument, among other projects, was considered by the large Telescope construction facility in Nanjing, created in 1958, and the potential of the Tibet plateau for a high-altitude optical station would lead in 1985 to a site survey.[27]

---

[26] In Beijing, I gave a seminar on infrared astronomy to the group of Hu Jing Yao, who had received a precious CCD detector given by an American colleague. At Beijing Normal University, cooperating with the observatory, an infrared detector (InSb), built in China, as the ones we were using in Chile for the study of stars in the infrared, was planned to equip the future 1.56-m telescope.

[27] Wang Shouguan, Lan Songzhu, Jiang Shiyang, Chen Hongqing, Qi Guangrong, Wang Zhenyi, Li Zhi, Liu Xinde. Preliminary investigation of the potential astronomical sites in Tibet, *Public. of the Shaanxi Astronomical Observatory*, 12, n°1-2, 1989.

A similar situation was occurring in radio astronomy with interferometric arrays. Finally, the access to astronomical space missions was essentially nonexistent, although a well-conceived infrared balloon program had just started in cooperation between Shanghai and Japan, a project considered of a very high priority by Ye Shuhua. Nanjing solar astronomers, with clever ideas about an X-ray imaging device, had plans for a satellite in the early 1980s. In fact, the instrument, adapted to a different scientific program, would be launched nearly 40 years later by the Long March 4B rocket, from the Taiyuan launch center.[28] Hence, the future of Chinese astronomy in space seemed very remote, although already a matter of thought and planning, of which the visit of a delegation to the French space agency, at the time of our visit, was a clear sign.

It was very interesting for us to assess, in each observatory, the presence of a new generation to carry astronomical research. As the universities had only reopened the entry exams in 1977, students were hardly present. The *zhi jing* (知青, the "educated youth")[29] came back progressively after 1974. Some, designated by the CAS to join a laboratory, had pursued a couple of academic years long ago but had no degree. In a sense "barefoot scientists," they were named "research trainees." In Kunming for instance, they were given a brief retraining with lectures on plasma physics and radio astronomy. The Cultural revolution had suppressed the national competition to become *aspirant-researcher*, created soon after 1949 for people aged less than 40 years and having an appropriate degree. This competition had just reopened in 1979, but no students were at the required level, and old staff members seized the new opportunity to get a better status and salary promotion. Nanjing, for instance, had 15 openings, but could only recruit four persons. After two years of advanced lectures and carrying research, and four additional years of research, the aspirant would pass a difficult national exam, and get the equivalent of a PhD. At this epoch, China did not have a formal PhD degree. The observatories, using these different ways, were in a growing phase, with typically planned recruitments of ten persons per year. It seemed to us unlikely that such growth could be effective, given the scarcity of financial resources to support new research projects.

In Shanghai, Ye Shuhua was our friendly host for several days (Fig. C4). Since 1976, the formal end of the Cultural Revolution following Mao's death, she had been re-appointed as research Professor at the Shanghai Observatory and would soon be elected to the Chinese Academy of Sciences (1980). She introduced us to the historical downtown station Zi Ka Wei (Xujiahui). Here stood the time service for China, funded by the Jesuits[30] in 1926 and her great former achievement for increased accuracy. Smiling, she took us to the garden, where roses were scarce at this season, and explained that during the Cultural revolution, she was protected

---

[28] *Launch of the world's first soft X-ray satellite with "Lobster-Eye" imaging technology.* https://phys.org/news/2020-07-world-soft-x-ray-satellite-lobster-eye.html

[29] The "educated youth" or *Shàngshān xiàxiāng yùndòng* (受过教育的青年) designates the millions of young Chinese sent to the countryside by the Cultural Revolution.

[30] Pierre Lejay, *L'Observatoire de Zi ka wei*, booklet in French, with photographs (1926).

Fig. C4 During the visit at Zi-Ka-Wei station in Shanghai, Ye Shuhua, Jean Delhaye and Wan Lai, Vice-Director. The board displays a friendly message: "*A warm welcome to the French astronomy delegation.*" (Photo: P.Léna)

from the Red Guards by Premier Zhou Enlai himself, who had ruled that China needed to keep an accurate time. Instead of being beaten and sent far away to work with peasants, she was given the relatively mild task to become the gardener of the observatory and could keep access to the international literature in this field.

With her, we visited the meridian instrument, installed in 1900 and an essential instrument for the time service, as well as a recent Danjon impersonal astrolabe, then the Zo Ze station, on a hilltop some twenty kilometers away from Shanghai. It was like plunging in the past since, founded by the French Jesuits, this station had a double equatorial refractor of 40 cm. A chapel was part of the station, kept intact in its Sulpician style of the 1900s. The new equipment was a satellite tracking laser, with a range of 2600 km, and a 6-m centimetric radio telescope, being part of the new Chinese VLBI network.

4. Emerging Cooperations

The 1979 mission, then the admission of mainland Chinese astronomy to IAU, opened new ways for active cooperation in astronomy since, for Chinese observatories to develop, large gaps would have to be bridged. French astronomers pursued in several directions the contacts made during the mission, especially with Ye Shuhua. Sending young scientists or engineers to Europe would be a top priority for the observatory directors of Kunming (Zhang Beirong), Nanjing, Beijing (Wang Shouguan) and especially Shanghai (Ye Shuhua). The XIVth IAU-UNESCO International School for young astronomers was held in Beijing in August 1986, with 42 Chinese participants, showing the emergence of a new generation of scientists eager to be trained.[31]

The Shanghai balloon-borne project, developed with Japan and strongly supported by Ye Shu hua, involved the young Jiang Dongrong who stayed in France in my group from 1981 to 1983. He could discover the CNES balloon launch facility in France (Aire sur l'Adour), then visit ESO in Germany. Yet, the necessary technologies of sensitive infrared detectors and cryogenics were for the moment outside the reach of China, and carefully controlled by the West for their military applications.

Planning for a future large telescope would become another hot issue in China,[32] which we had felt during our visit, with site testing initiated in Tibet.

To conclude, our visit showed, once more, that science and scientists can play a significant role, especially at times of political tensions, to build a world of joint actions and mutual understanding.

---

[31] To contribute to this training, Prof. Hu Jing Yao, whom I had met in Beijing Observatory, joined by Sun Weixin from Taipei. They translated into Chinese my own textbook for graduate studies, *Observational methods in astrophysics*, which was published in Taiwan in 2004, and as a new edition in Beijing in 2015.

[32] Letter of P. Léna (December 16, 1987) to Ye Shuhua, a few days after the formal approval of the VLT by ESO Council, inviting her to send someone to the forthcoming ESO-NOAO Conference and the Large Telescope Conference in Germany. She immediately replied (January 25, 1988) and sent to Germany her Vice-director and chief engineer, Zhu Nenghong.

Appendix D: *Under One Starry Sky*—The History of the International Astronomy Union (IAU)

Written and edited by Fu Chengqi and Ye Shuhua[33]
  (Shanghai Observatory)
  Published by the Shanghai Jiao Tong University Press (2009)
  (中国上海交通大学出版社)

Chapter 5: China and the IAU
4. The contraflow of history
5. The return to the IAU

Translated by Yi Zhou, Institut d'Etudes Politiques de Paris[34]; approved by author Ye Shuhua, with permission from the publisher.

Note: The book by Fu & Ye, entitled 《同一个星空国际天文学联合会史》 "*Under one starry sky* (etc.)" is a history of modern astronomy in China, from the early contacts with Western astronomers and Jesuit priests to contemporary times. Chapter 5 covers the topic "*China and the IAU*" (starting from 1935, when China first joined the IAU, until now).

The whole book has unfortunately not been translated, but we present in this Appendix (with the kind permission from the authors and the publisher) the translation of Sections 4 & 5 of Chapter 5 that concern the period in which China withdrew from the IAU in 1958 until rejoining the Union in 1982. (The full translation of Chapter 5 is now part of the IAU archives, but can be obtained upon request from the authors.)

---

[33]IAU Vice-President (1988–1994). [The authors' family names are "Fu" and "Ye"].

[34]Widely known as "Sciences Po." Graduate of its Paris School of Interantional Affairs [last name Zhou (周), first name Yi (易)].

T. Montmerle, Y. Zhou, *China and the International Astronomical Union*, Historical & Cultural Astronomy, https://doi.org/10.1007/978-3-031-01787-2

Chapter 5, Sect. 4: The Contraflow of History

Soon after it became a member of the IAU in 1935, China experienced a war against the Japanese invaders that lasted eight years—the Second Sino-Japanese War. The *Academia Sinica Institute of Astronomy* and its Purple Mountain Observatory in Nanjing, as well as the Chinese Astronomical Society, all had to retreat to the Southwest of China, far away from Japanese occupation. The war had made the Chinese Astronomical Society unable and too distracted from its participation in IAU's activities. It stopped sending its members to attend the IAU General Assemblies after 1935. The only Chinese person who participated at the IAU's sixth General Assembly in Stockholm was Dai Wensai (戴文赛). Dai was a foreign student studying in the UK at the time. After the Stockholm General Assembly, it seemed that the Chinese annual contribution to the IAU only lasted two years, 1938 and 1939. As a member country, China had to pay 300 Francs[35] to the IAU each year. There does not exist any earlier records that show China has been paying its dues to the IAU previously. It was not until 1947, two years after the war that the Chinese Astronomical Society was finally able to reimburse the IAU eight years of membership fees, a total amount of 1568 dollars. (600 Francs each year, during that time 1 Franc was roughly equivalent of 0.3266 dollar.) In the meantime, the Chinese individual memberships to the IAU also drastically diminished. The most individual members of the IAU that China had was in 1938, there were 11 astronomers (not including the French Jesuit Pierre Lejay or 雁月飞 in Chinese. He was working at the Xujiahui Observatory in Shanghai). Besides the earliest Chinese members of the IAU, such as Yu Qingsong(余青松), Gao Lu (高鲁), and Jiang Bingran (蒋丙然), there were also 8 new members by this time. They were Zhang Yuzhe (张钰哲), Chen Zungui (陈遵妫), Gao Pingzi (高平子), Li Heng (李珩, Zhao Jinyi (赵进), Zhu Wenxin (朱文鑫), Zhang Yun (张云), and Shen Xuan (沈璿).

Among them, Zhu Wenxin (1883–1938) studied at the University of Wisconsin between 1907 and 1910. He later became a professor at the Nanyang University and Fudan University in Shanghai. However, there were only two individual Chinese IAU members left in 1948. They were Zhang Yuzhe of the Nanjing Astronomy Institute, and Jiang Bingran of the National Taiwan University. On top of that, the Chinese Astronomical Society paid its membership fees in 1948 for a total amount of 196 dollars. Its link with the IAU was then once again interrupted by a new war, which was the Chinese civil war fought between the Nationalists and the Communists. The year of 1948 was during the height of the Chinese Civil War, the Chinese Astronomical Society was financially supported by the Nationalist government. Thus, the Chinese Astronomical Society was unable to participate in the IAU's activities during wartime. The Society did not send its members to attend the seventh IAU General Assembly in Zurich. Cheng Maolan (程茂兰, 1905–1978), who was

---

[35] At the time, the "unit of contribution" to the IAU was expressed in Gold Francs (Blaauw 1994, p. 258).

working at the Lyon Observatory at the time, participated in this General Assembly as an individual French member of the IAU.

After the founding of the People's Republic of China (PRC) in 1949, it created an immense future for the development of astronomy in Mainland China. It was only natural that the PRC would inherit the Chinese membership in the IAU. In the early days of the PRC, all was ready to be undertaken. During the Korean War, Western countries had adopted a strategy of blockade, isolation, and subversion vis-à-vis Mainland China. There was also the Sovietization of the entire Chinese higher education and research system since 1951. These all resulted in that the Chinese Astronomical Society was unable to resume its participation within the IAU. It did not send any delegates to attend the IAU's Eighth General Assembly in Rome in 1952. Furthermore, the Chinese Astronomical Society did not pay its dues to the IAU between 1949 and 1954. It was not until 1955 that the Chinese Mainland resumed paying its membership fees to the IAU. The annual amount was 1000 francs. The Society was able to repay the entire six years of membership fees that were due. It was an accumulated amount of 6000 francs (around 1,959 dollars according to the exchange rate at the time). Moreover, Chinese Mainland astronomers also attended the Ninth IAU General Assembly that year in Dublin. They were Zhang Yuzhe, Dai Wensai, Wu Xinmou (吴新谋, 1910–1989), and Ye Shihui (叶式辉). Zhang Yuzhe even served as a member of the IAU Finance Committee during this General Assembly. Three years later, seven Chinese Mainland astronomers represented the Chinese Astronomical Society and attended the 10th IAU General Assembly in Moscow. This delegation included Zhang Yuzhe, Cheng Maolan, Zhao Quemin (赵却民), Zhu Renjun (朱人俊, 1916–1993), Li Heng, Dai Wensai, Wang Shouguan (王绶琯). This delegation also invited Ye Shihui, who was studying in the USSR as their translator.

There were two main reasons for which these Chinese Mainland astronomers were able to return to the IAU in 1955. On the one hand, there was not a lot of obstruction for them to participate in this General Assembly. It was only reasonable for the Chinese Mainland's delegation to inherit the membership of the former Nanjing's Chinese Astronomical Society. It was also important to know that the entity that was registered at the IAU was Nanjing's Chinese Astronomical Society: The international astronomy community recognized the Chinese Mainland, not Taiwan.

On the other hand, the participation of Chinese Mainland astronomers at the Dublin General Assembly was the result of various matured conditions. Starting from 1948, the IAU Executive Committee finally had a member from the Socialist Bloc. The celebrated Soviet astronomer Viktor Ambartsumian took over as the Vice-President of the IAU. He recommended to China to resume its participation in the IAU's General Assemblies. Furthermore, the Korean War ended in 1954. The revival of the Chinese economy and the consolidation of power in China gave the Chinese government the necessary ability to pioneer in different civil foreign interactions. The Chinese government also gained the financial capacity to repay all the due membership fees that were neglected in the previous years. As recalled by the former President of the IAU Leo Goldberg (1913–1987) at the 16th IAU General

Assembly in Grenoble, France, he suggested that the restoring of the Chinese membership was the wish of the then President of the IAU, Otto Struve. Struve had sent an invitation letter to Zhang Yuzhe, who was the President of the Chinese Astronomical Society in 1955. In all, there was not much obstruction in resuming the Chinese Mainland's membership in the IAU. However, what came as a surprise for the Mainland astronomers was that, after their return to the IAU, there was a historical turnaround which happened at the Moscow General Assembly, three years later. A small number of countries, led by the United States, staged a farce to exclude Chinese Mainland astronomers through Taiwan's membership.

The establishment of the "New China" plunged the U.S. government into a state of powerlessness and sadness. This was clearly demonstrated by the White Paper on "*United States Relations with China*" released by the U.S. Department of State in 1949. The then U.S. Secretary of State Dean Acheson further proved this sentiment in his letter exchanges with the then President Harry Truman. In which, Acheson stated that "The unfortunate but inescapable fact is that the ominous result of the civil war in China was beyond the control of the government of the United States." Therefore, the U.S. government had adopted a hostile policy toward the New China from the beginning. On one side, it systematically propped up the Chiang Kai-Shek's clique by supporting and defending Taiwan economically, militarily, and politically. After the Korean War broke out, it became a naked excuse of intervention by the Americans in interfering the cross-Strait relations. On June 27, 1950, the United States Seventh Fleet invaded the Taiwan Strait. The Thirteenth Air Force was then stationed in Taiwan. In December 1954, the U.S. signed the "(Sino-American, omit in the Chinese original text) Mutual Defense Treaty" with Chiang Kai-Shek's clique. This development further aggravated the long-term standoff and tensions across the Taiwan Strait. Ever since then, the Taiwan issue has become a major dispute between the USA and China.

On the other side, the U.S. government carried out an isolation, containment, blockade, bullying, and obstructing policy toward the New China. However, as the consolidation of the regime and the growth of the economy progressed, China's international stranding was able to restore quickly. In 1956, 25 countries had already established diplomatic relations with China. The failure of its isolation strategy forced the U.S. government to change their policy toward China. They subsequently adopted the tactic of "*two Chinas*," "*one China, one Taiwan*." Washington abetted Taiwan to participate in all kinds of international governmental and civil organizations as one of the meanings to implement this policy. Thus, the IAU certainly became one of its targets.

In the early 1950s, Taiwan did not participate in any of the IAU's activities. It was not a surprise at all that Chiang Kai-Shek's Nationalists did not care for any sciences activities after their recent retreat to Taiwan. Nonetheless, Taiwan had suddenly emerged as a problem in the late 1950s. This had come as a surprise. The Taiwan issue had first appeared to the Executive Committee of the IAU in 1957. On July 4 that year, the Executive Committee had a meeting in Liège, Belgium. The former IAU President who served as an advisor at that time, Otto Struve, was an American astronomer of Russian descent. On his way to attend this meeting, he passed by the

American Embassy in Brussels. It was obvious that he had very likely discussed some political issues regarding the IAU with the American officials. For example, they could have discussed the soon to be held IAU General Assembly in Moscow (1958) and the American invitation process of the next General Assembly in 1961. The order that Otto Struve received was that he had to take in consideration of the U.S. government's policy toward Taiwan. In other words, they had to insist that Taiwan was the legitimate representative of China in any international conferences. Therefore, the IAU had no choice but to invite Taiwan to its General Assemblies, in spite of the fact that it was not a member of the IAU. As to what extent that the Executive Committee had discussed this issue was unknown to us. No details are available about this meeting. We only knew that one of the issues discussed at this meeting was the 1961 General Assembly. Those who participated this meeting in Liège included all the members of the IAU Executive Committee at the time, and Otto Struve as an advisor. In addition, the Russian astronomer P. G. Kulikovsky attended this meeting as a guest. She was working at the Sternberg Astronomical Institute in Moscow at the time. However, the activities and correspondence among Struve, Goldberg, the United States National Academy of Sciences (NAS), and the U.S. Department of State after this meeting in 1957 and before the Moscow GA in 1958 gives us some additional clues. It is evident that the U.S. Department of State forced the IAU to accept Taiwan's participation by threatening to deny visas to astronomers from socialist countries and by renouncing the American invitations. At the same time, Struve and Goldberg's refusal in mingling the Taiwan problem and the U.S.'s invitation process of the 1961 GA was well documented. The NAS backed their position on this issue as well.

As early as 1957, in fear of U.S. government could potentially deny the visa applications of certain attendees, Leo Goldberg went to discuss this matter with the U.S. Department of State. He was serving as the President of the U.S. National Committee for the IAU at the time (USNC/IAU). Goldberg was also accompanied by Detlev Bronk (1897–1975), who was the President of the NAS at the time of this trip. They tried to obtain a guarantee from the State Department that it would grant visas to all applicants that were invited to the Berkeley GA. In November that year, the NAS again voted for calling for the State Department to send out invitations following the IAU guidelines. From Goldberg's letter, in an exchange with Struve dated January 8, 1958, it was not difficult to see their efforts in trying to convince the U.S. government and Washington was reluctant to do so:

> The U.S. Department of State is quite clear on this issue at the Deputy Secretary of State level... Since last summer we are continuing to fight for this idea.

It was evident that the politicians in the U.S. Department of State used their visa power to play the Taiwan card, regardless of losing the reputation of science studies in the USA. In March 1958, both Goldberg and Struve extensively discussed with the State Department the idea of holding an international science conference in the USA. During their discussions, it was obvious that the State Department asked the President of the IAU to welcome Taiwan as a member. This was distinctively reflected in the letter written by Goldberg to the U.S. Department of State on March 24 that year.

(I strongly) oppose the idea of abandoning our invitations unless inviting Formosa (Taiwan) to become a member... I believe any attempts of combining the invitation to the Nationalist Chinese and the U.S.' General invitations to the General Assembly would possibly considered by the IAU's Executive Committee as political interference. I think this will damage the reputation of sciences in the USA.

Goldberg further wrote,

I am more leaning towards conducting some investigations about the current state of Formosa's astronomy studies. If their situation is satisfying, I will urge to restore its IAU membership. However, it is completely irrelevant to put it as the prerequisite condition to invite the IAU to have their General Assembly in the USA, at least from the point of view of sciences. . . .Should this meeting be held in some other places, the IAU, before its Moscow General Assembly, has sufficient time to promote another suitable host country to send out the invitations. . .

Goldberg's letters completely exposed the sudden emergence of the Taiwan issue in the IAU Moscow General Assembly and later on, this was the result of the U.S. government's planning, interference, and abatement.

Editor's note: One of the major issues on the agenda of the U.S. National Committee to the IAU in the mid-1950s was to gain State Department clearance to invite the Union, the world body of astronomy, to hold its next General Assembly in the United States in 1961. The issue was that the Immigration and Nationality Act of 1952 (INA, also known as the McCarran–Walter Act), could severely curtail travel to America by foreigners from suspect countries, and even those who were outspoken from friendly countries. In the general reactionary political atmosphere of the day, this resulted in numerous difficulties for participation in international scientific activities in the USA. For instance, both China and Russia had been members of the IAU since July 1935, but in the Cold War world of the 1950s, and the creation of the 'two Chinas', the U.S. Department of State demanded that Nationalist China, what is now Taiwan (on the island of Formosa) be invited and 'Red China' excluded. Even though the former did not have an active astronomical community and the latter certainly did. *Here we retrace the challenges faced by IAU officers ranging from Leo Goldberg,* Adriaan Blaauw, and Patrick Wayman among others to finally recognize and welcome the PRC, while retaining Taiwan as a separate member. Through many efforts stimulated by Cold War pressures, Taiwan by the 1980s had grown a vibrant astronomical community.

On April 30, 1958, before the Moscow General Assembly, a month after Goldberg had written to the U.S. Department of State, Taiwan, had handed in its application to the IAU to apply for membership status. This letter was written under the name of the "President of the Astronomical Society of the Republic of China (Taiwan)" to the Executive Committee of the IAU. It roughly mentioned that the above Astronomical Society has officially started to work with Taiwan's Academia Sinica since early April that year. Because Academia Sinica is a member of the International Council of Scientific Unions (ICSU); therefore, "the Astronomical Society of the Republic of China (Taiwan) or ASROC" should naturally have a membership in the IAU. It was their intention to officially to apply as well. Soon after, the Executive Committee received a letter from the Academia Sinica dated May 5. This letter had proven Taiwan's official application intent. Since then, the

Taiwan issue became public within the IAU. However, the Taiwan side did not understand the relationship between the ICSU and the IAU. They thought by joining the ICSU, they would be qualified to enter the IAU as a member. In fact, as early as 1931, after ICSU has replaced its former body the International Research Council (IRC), it has changed its rules and no longer required the IAU member countries to be ICSU members as well. Those countries that are not members of the ICSU could also join the IAU if the Executive Committee approved their applications. However, the Statutes of the IAU clearly states that, one country can only have one astronomical representation as a member. Consequently, when Nanjing's Chinese Astronomical Society had already represented China, accepting Taiwan's application to the IAU was an obvious violation of its Statutes. This incident has provoked the "two Chinas" dispute. In fact, from the legal point of view, there was no such a thing as "the Astronomical Society of the Republic of China (Taiwan)". Taiwan authorities' "Ministry of Interior" only approved its status as a civil entity only one month prior to the Moscow General Assembly. This was in July 1958 and was at least two months later than the deadline for application to the IAU. "The Astronomical Society of the Republic of China (Taiwan)" was legally registered in Taipei. Its President was Jiang Bingran. Jiang had left Mainland China for Taiwan in 1946. Its establishment was so hasty and its application to the IAU was rather impatient as well. These circumstantial evidences all proved that the main instigator hiding behind must have been the U.S. government at the time.

The then General Secretary of the IAU, the Dutch astronomer Pieter Oosterhoff (1904–1978) was extremely busy with the preparation of the Moscow General Assembly. Therefore, under such circumstances, it was impossible to discuss Taiwan's adhesion to the IAU in such a hurry. Instead, the decision was they would postpone the relevant discussion, in order to enable the newly elected Executive Committee to require further information and then proceed to debates and decision-making. On May 19, Oosterhoff informed President Jiang Bingran that, Taiwan's application to the IAU must follow the instructions laid out by the IAU Statutes. As a result, Taiwan's application would be decided at the earliest at the 1961 IAU General Assembly.

The Chinese Mainland delegation arrived in Moscow on August 9, 1958. Soon after their arrival in the USSR capital, they had received one good news and one bad news from the then Vice-President of the IAU, the Russian astronomer Boris Kukarkin (1909–1977), and Viktor Ambartsumian. The bad news was that there was an application from Taiwan, and the good news was that the Socialist Bloc had recommended the Mainland astronomer Cheng Maolan to become the Vice-President of the IAU. He was serving as the Director of the Office of Preparation of the Beijing Observatory at the time.

Under the Cold War political context in the early 1950s, the diametrically opposed two blocs recommended the IAU Executive Committee's members. Each bloc could recommend 2–3 candidates, if no opposition was risen by the other side, then his/her nomination would be approved. Therefore, Cheng Maolan was an excellent candidate to represent Mainland China, because he just recently returned from France in 1957. Cheng started to study and work in France as early as 1925. He had received his Bachelor's, Master's, and PhD all from universities in France. Cheng earned his Master's degree from L'Université de Lyon in Mathematics with a scholarship from the French National Centre for Scientific Research (CNRS). He then studied astrophysics under the French astronomer Jean Dufay and obtained his PhD in Mathematics from the CNRS in 1939. Having received his PhD, Cheng Maolan subsequently conducted researches at the L'Observatoire de Lyon and L'Observatoire de Haute-Provence. In 1949, he became a senior research supervisor at the CNRS. Cheng also received a Légion d'Honneur from the French Ministry of Education.[36] Since Cheng had lived in France for 32 years, he had accumulated strong friendships with the French astronomers. This special relationship would reduce many of the obstacles for Cheng's nomination; this was among the many considerations that the Socialist bloc recommended him.

After learning these two major pieces of information, the Chinese delegation treated them with utmost importance. The Deputy Director of the Purple Mountain Observatory—Zhu Renjun (朱人俊) quickly reported this news to the Chinese ambassador in Moscow—Liao Xiao (刘晓). As the Deputy Party branch Secretary at the Purple Mountain Observatory, Zhu was also in charge of the Party affairs of this Chinese astronomy delegation. Soon after, the Chinese embassy in Moscow received orders from the Chinese Ministry of Foreign Affairs in Beijing. Beijing ordered them to firmly oppose the invitation of Taiwan and to give up the nomination to the vice-presidency of the IAU. The first order came as expected; however, the second one was rather impossible to comprehend. It caught all of the members of the Chinese delegation by surprise. They assumed if they could have a Mainland scientist at the table of the Executive Committee would be beneficial to solve the Taiwan issue. As one of the delegates Mr. Wang Shouguan (王绥琯) recalled, all members from the Mainland delegation were largely disappointed by this order, so was Kukarkin himself. The head of the delegation, Zhang Yuzhe who normally was rather a reserved person, even expressed strong reactions. After more than half of the century, maybe in retrospect, this was the right decision. Politics was deeply involved and played a huge role in academic organizations, as the two opposing blocs were already right against each other. Under these complex circumstances, Cheng Maolan would most likely not be able to address such a sensitive issue single-handedly. He just came back to China recently. Besides, there were many other risks as well; it was a better idea to give up his nomination.

After having received orders from the Chinese embassy, the Chinese delegations prepared for possible scenarios prior to the General Assembly. In the worst-case scenario, they were ready to withdraw from the Moscow GA completely, if the Taiwan issue was ever raised. However, this was not the case.

On August 19, the IAU Executive Committee held a meeting, under the direction of the newly elected President Jan H. Oort, and it took the decision that the Taiwan membership issue would be pushed back to be discussed in the next meeting. The

---

[36] It was actually the "Palmes Académiques" awarded by the Ministry of Education. For a detailed biography of Cheng Maolan, see MZG22, *The two lives of Cheng Maolan* (Springer), in press.

Executive Committee also required more information on Taiwan's astronomical activities. They sent a letter to the President of Taiwan's Academia Sinica to explain this situation in the name of the soon-to-be retired IAU President André Danjon. In this letter dated August 15, Danjon reiterated the opinion expressed by Pieter Oosterhoff previously. He said that it was a total misunderstanding to think that they would automatically acquire membership at the IAU being a member of the ICSU. The letter further explained that it was also a misunderstanding to think the Taiwan membership question would be resolved during the Moscow GA. In additionally, Danjon raised some questions about the level of astronomical activities in Taiwan. For example, he pointed out that the IAU did not know where Taiwan's observatories were located.

However, Danjon promised that the IAU Executive Committee would reconsider Taiwan's application in 1959. Danjon did not mention that this meeting would discuss Taiwan's membership, nor he outright rejected this possibility. He agreed the Taiwan issue would be debated at the next meeting. This reflected its difficulty and the division among the Executive Committee. Because of the high sensitivity of the Taiwan issue, neither the working report of the 1958 Executive Committee nor that General Assembly's official publication in 1958 mentioned Taiwan's application for a membership at the IAU.

Taiwan's application to the IAU was rather quietly treated in Moscow, the main reason being specifically explained through Leo Goldberg's report to the President of the NAS. Goldberg served as the head of the U.S. astronomy delegation, as well as the Director of the University of Michigan Observatory at the time. Goldberg pointed out in his letter that, if the IAU accepted the application from Taiwan, it would have significant consequences. He suggested that not only Mainland China would withdraw from the IAU, but that the Soviets may follow suit as well. The latter reason might have been that Taiwan did not join the IAU in Moscow as it had wished. It was unimaginable to foresee what an awkward situation that would have been if the host country decided to pull out from the General Assembly. Many years later, when the Belgian astronomer Pol Swings made a speech when he left the position of President of the IAU, Swings quoted Danjon as the latter having said that, "exercising the President's role was as an aviator always on high alert so that he won't drive the plane into a cliff." This depiction vividly described Jan H. Oort's state of mind, as he was the President of the Executive Committee at the time.

Nevertheless, what happened next all changed drastically. The next Executive Committee meeting took place in September 1959. It extensively discussed Taiwan's application. The then General Secretary of the IAU Donald H. Sadler reported his interactions with Taiwan to the Committee. He put forward the idea that Taiwan's application was generally in accordance with the Statutes of the IAU. Kukarkin and the Czechoslovakia astronomer Bohumil Šternberk (1897–1983), who were both Vice-Presidents of the IAU, firmly opposed Taiwan's integration into the IAU. They argued that the level of academic research should be the only material in reviewing an application to the Union. They suggested Taiwan's astronomical activities were not independent, thus making them unfit to apply for its membership. Since it is stated in the third clause of this Statutes that, the definition of "countries" indicate "any autonomous region, protectorates and territories that have independent astronomical activities." (This is not what is written in *the current Statutes of the IAU.*) Thus, the requirement of having independent astronomical activities was counted as essential for membership. Besides, Kukarkin also argued that if the Executive Committee was to accept Taiwan's application, the Mainland would withdraw from the IAU. After some fierce debates, the Executive Committee passed the motion that it would accept Taiwan's application by a vote of 5 versus 2. Clearly, it was highly probable that the other Vice-Presidents from the USA, West Germany, Canada and South Africa all voted in favor of accepting Taiwan's application. Sadler who was from the UK likely also voted in favor as well.

After this meeting, Kukarkin issued a statement on behalf of those member countries that opposed Taiwan's application. He reiterated his arguments in this letter and argued to maintain the status quo until the next General Assembly in 1961. Still, the Executive Committee insisted upon its decision. Their reason was that the Statues of the IAU did not include clauses to reject an application.

This decision quickly enraged many parties. The President of the Chinese Astronomical Society Zhang Yuzhe expressed his protest against the Executive Committee of the IAU. On his letter dated on November 20, 1959, Zhang clearly argued to the President of the IAU Jan H. Oort that, "Taiwan is an inalienable part of the Chinese territory; it is a province of the PRC." He stated if the content of the report was real, "I would then represent the PRC to express my strong protest to you." Zhang also urged the Executive Committee to abolish this illegal decision. He stated ". . .or else, our Astronomical Society would. . . firmly withdraw from the IAU." Unfortunately, Oort would not back down. In his response letter dated on December 2, 1959, Oort once again emphasized the so-called reasons that "The Republic of China (Taiwan, Taipei)'s application to the IAU was accepted on the grounds provided by the Chapter 2, Clause 3b of the Statutes." It stated that "the applications. . . should be in accordance with the Statutes of the IAU in form, as well as the Statutes and Rules of Procedure of the ICSU." Oort stressed that "because The Republic of China (Taiwan)'s Astronomical Society has already received the agreement from the IAU, the Executive Committee's decision is therefore unchangeable." He once again highlighted the idea of "two Chinas" in his letter. Oort pointed out that "Taipei Astronomical Society only represents that of Taiwan, its leadership, management, and representation are all independent from the Chinese Astronomical Society in Beijing." He asked the Chinese Astronomical Society to reconsider their point of view based on this letter. The Clause 3b that Oort mentioned was one of the requirements stated for countries to apply for a membership in the IAU. It stated that "An ICSU member country can join the IAU by integrating into ICSU sub-organizations or a recognized national astronomical committee." Evidently, Oort not only forgot he mentioned the need to investigate the status of Taiwan astronomical activities, but also did not reinstate the importance of "independent astronomical activities" being among the requirements of joining the IAU.

Having received the letter of protest from the Chinese Astronomical Society, the General Secretary of the IAU Donald H. Sadler instinctively made the decision to publish the Taiwan membership vote result on the second issue of the IAU's information Bulletin in November 1959. The news of Taiwan's admittance to the IAU was included in this publication as a fait accompli. This decision made by the Executive Committee of the IAU was irreversible. On February 5, 1960, Zhang Yuzhe once again expressed his opposition and protest to the President Jan H. Oort, in the name of the President of the Chinese Astronomical Society of the PRC. Zhang expressed that the Mainland China's participation would be halt immediately and it formally withdrew from the IAU and all of its sub-committees. Zhang Yuzhe also stated all relations with the IAU was therefore broken completely. At the same time, all individual Mainland members of the IAU issued their personal statements and declared their resignation from the IAU. The official Xinhua News Agency of the PRC also published an article titled "A protestation to the fabrication of the two Chinas plot." It officially announced the PRC's withdraw from the IAU.

From then on, the PRC had left the IAU and severed all ties with the latter body. Soon after, the 3rd edition of the IAU Information Bulletin in 1960 printed the resignation of the Chinese Mainland from the IUA on its front page. It was written that "the President of the PRC's Astronomical Society, Professor Zhang Yuzhe officially announced that the PRC has left the IAU. This is a gesture of protest against the Executive Committee's decision in accepting Taiwan (the ROC) as one of its members. . . From this day and onwards, the PRC astronomers are no longer members of the IAU, therefore nor would they still remain as members of its professional committees."

This faulty decision taken by the Executive Committee of the IAU had caused a wave of indignation from the Polish Academy of Sciences and the Bulgarian Academy of Sciences. They all demanded the Executive Committee to reconsider. During the Executive Committee meeting in July 1960, Sadler refused this request in his working report. He stressed, "the reversal of prior decisions would not work." He argued "only the General Assembly can overwrite the decision taken by the Executive Committee through a majority vote. . ." He also advised all professional committees to take off the Chinese scientists' names from their name lists.

The forced exile of the Chinese Astronomical Society away from the IAU was the pure result of the isolation and hostile policy adopted by the U.S. administrations since 1949. Furthermore, this may also had to do with the 1961 IAU General Assembly. American astronomers had always expressed their strong longing of organizing a General Assembly on their home soil. The then American Vice-President of the IAU Leo Goldberg and his predecessor Otto Struve had all expressed their wish on multiple occasions.

Nonetheless, one of the factual impediments that they were forced to face was that the American Astronomical Society must act as a sponsor so that all invitations would be approved by the U.S. Department of State. Only with the approval of the U.S. government for their visas, all of the foreign invited parties can then attend this General Assembly. This is the most basic requirement from the IAU for any of the host countries, this type of sponsorship should include all member countries and Mainland China was not an exception. Yet, this simple demand was against the U.S. government's isolationist policy toward the PRC. Obstructing the Chinese Mainland scientists to participate in this General Assembly was the only remedy for the U.S. government to overcome this intricate situation. Their approval for Taiwan to join the 1961 General Assembly could then kill three birds with one stone. This solution not only followed the wishes of the U.S. government but also outcast Mainland China and push or Taiwan's integration in the IAU. This gesture once again demonstrated its twisted evil face during the Cold War.

While the IAU General Assembly was held in Berkeley California in 1961, the Czechoslovak and Soviet Union Academy of Sciences each put forward one resolution in regards to the Taiwan issue. The Czechoslovak side asked "the General Assembly to publicly declare the nullification of the decision taken by the Executive Committee in accepting Taiwan's membership in September 1959." The Soviet proposal was even more precise, it indicated this decision "had seriously damaged the reputation and the interests of the IAU, as well as harmed the international astronomical cooperation." The Soviets once again pointed out Taiwan did not have "independent astronomical activities." They pushed forward the idea that "the Executive Committee's decision was in opposition to the Statues and spirits of the IAU." They required the General Assembly to abolish the Taiwan decision. These two resolutions were combined into resolution No. 10 (a) to the General Assembly that year. Once the Berkeley General Assembly started, the General Secretary of the IAU announced that the great majority of the Executive Committee members had opposing views on those that had been expressed in this resolution. Soon after, despite the opposition from the Soviet and Czechoslovak delegations, President Oort read out an already prepared statement from the Executive Committee on this matter.

Oort proclaimed that "I, as the representative of the Executive Committee of the IAU oppose this resolution and ask all member countries to vote against it."

As the President of the IAU, the way Jan H. Oort publicly expressed his rejection against a resolution, and he prompted others to follow suit was extremely rare and impolite. In his following remarks, Oort completely overlooked the rapid development of astronomy in the Chinese Mainland. He also once again boasted the idea that "the Executive Committee would not and could not reject Taiwan's application." Oort stated, "Taiwan has the right to join our Union." He claimed that "although the astronomical activities in Taiwan were small, they went in the same direction as the IAU's policy of promoting the development of astronomy in this type of countries" Oort stubbornly shoved down the idea that "Taiwan was already a member of ISCU, whom the IAU was also a part of." He reasoned that "no proofs had demonstrated the astronomical activities in Taiwan were not independent. It followed guidelines from the self-formed national astronomical committee." President Oort put forward the idea that there was a need to learn more about Taiwan's astronomical activities and being a member of the ICSU did not necessarily translate into an automatic membership at the IAU. Now, he did not care about the Statutes of the IAU anymore. On the contrary, he believed it was a "unilateral decision by the Chinese" to leave the IAU and to break all ties. He egregiously knew that if "Taiwan's application was to be accepted, the Chinese Mainland would leave the IAU," yet he still pushed for the illusion that "we still want the PRC to continue to join the Union." Before he once again boasted the decision made by the Executive Committee, Oort committed yet another faulty elaboration.

"In the light that the Two China Republics can be mixed up in various occasions, and to distinctly indicate the two geographic regions that conduct independent astronomical activities, we decided our Union would therefore use the following acronyms to describe these two countries in any meetings, only excluding official letter and documents exchange":

1.	
Official Name:	The People's Republic of China
The location of government:	Beijing
Membership organization:	Nanjing National Astronomy Committee
Membership since:	1935
Geographic area of representing astronomers:	Chinese Mainland
Acronym name:	China
2.	
Official Name:	The Republic of China
The location of government:	Taipei, Taiwan
Membership organization:	The Astronomical Society of the Republic of China (Taipei)
Membership since:	1959
Geographic area of representing astronomers:	Taiwan (Formosa)
Acronym name:	Taiwan

Under the rule of the few, the Berkeley General Assembly did not take enough time to discuss the Soviet resolution and subsequently voted in a hasty manner. Although Kukarkin denounced this Executive Committee statement had nothing to do with him, yet under such a circumstance, the final result was rather predictable. On the voting day (August 24, 1961), the Czechoslovakia–Soviet Union's resolution received five votes in favor, twenty-four against and four abstentions. This resolution was rejected.

The Berkeley General Assembly's results of "Two Chinas," "One China, One Taiwan" were naturally opposed by both sides of the Taiwan Strait. Just as foreseen by General Secretary Sadler, the Executive Committee had received a letter from Academia Sinica in Taipei, which stated they were also against the committee's decision. The Academia Sinica asked the Executive Committee to forward Taiwan's official name question to the General Assembly. We do not know the exact content of this letter, but based on the notes taken during this General Assembly, the Taiwanese side also expressed their disapproval of the "Two Chinas" proposal. Even so, the General Assembly did not discuss the name issue as Taiwan had wished. General Secretary Sadler explained that they had received this letter too late, so it was not included in the agenda of the Berkeley General Assembly. He stated unless there were more than half the member countries agreed to this idea, the General Assembly would not discuss on this matter.

The rejection of the Czechoslovakia–Soviet Union resolution had also clearly to do with other matters as well. Notes from the Executive Committee on August 14 mentioned clearly that, if this resolution would be passed,

> Most members including the President, the General Secretary of the IAU thought their resignation was then inevitable, but they also knew that this should not be used as a threat...

The possibility of an incoming crisis of the Executive Committee may have motivated member countries to vote against this resolution. In fact, during the Berkeley General Assembly, the West German member of the Executive Committee, Otto Heckmann (1901–1983) once expressed his doubts over the membership debate. He pointed out that "the IAU's policies towards individual members and national members were contradictory." Heckmann suggested that "the Statues of the IAU should be modified so that it would allow the Executive Committee to reject applications..." However, Sadler dismissed his recommendation, as "it would be challenging to actually carry out such suggestions." Yet, General Secretary Sadler never actually described where the challenges would come from. Maybe it was due to a series of unusual events, this Executive Committee meeting only briefly mentioned Taiwan issue. It stated since the last General Assembly, Taiwan had become a member of the IAU, and China left this organization.

The rejection of the Czechoslovakia–Soviet Union resolution made China the first country to quit the IAU due to political reasons, not because of financial difficulties. In the next nearly 20 years, China had broken off completely with the IAU. What happened between 1958 and 1960 within the IAU over the Taiwan issue was all but the fault of the U.S. government. In the 1972 edition of the NAS "Biographical Memoirs," it was stated in the Volumes 72 on the life of Leo Goldberg that:

> The cold-war rivalries in the United States and the Soviet Union flourished in the 1950s. The Soviet Union invited the IAU to convene in Moscow in 1958 and proclaimed that astronomers from all member countries would be welcome. To save face, the United States would have to host the meetings in 1961 under the same guarantees that the Soviet Union had given. Astronomers from all member countries would be welcome. In particular, since the People's Republic of China (PRC) was an IAU member, its astronomers would be allowed to attend, but Taiwan was not a member at that time. PRC athletes were to attend the 1960 winter Olympic Games in the United States. The Department of State headed by Secretary Dulles observed that this action was ad hoc and did not imply that mere scientists could expect such favors. Goldberg contacted his representative in the U.S. Congress, George Meader, a conservative and fair-minded Republican, who presented the case to Dulles, who referred it to his science advisor Wallace Brode. Brode promptly demanded that Taiwan be invited to the IAU.

> The fact that Taiwan then had no astronomers and would have to qualify for IAU membership in the approved way meant nothing to the militant anti-Communist Brode. Brode wanted Goldberg to go to the 1958 Moscow meeting and submit the 1961 invitation but with the condition that Taiwan be admitted at once. Such a demand could well wreck the IAU. From Brode's point of view, if the astronomers would not go along with his orders, so much the worse for them.

> Goldberg was unable to submit to these demands and offered to resign, but the National Academy of Sciences supported Goldberg's position that the IAU should act on the

application of Taiwan in an orderly way. The invitation for the IAU to meet in the United States was issued and accepted. Taiwan was admitted to membership in 1959; the PRC withdrew in 1960, but returned later.

This passage and the events that it described could not serve as better proofs of our argument.

Chapter 5, Sect. 5: The Return to the IAU

After the Berkeley General Assembly in 1961, the Chinese Mainland astronomers ceased to participate in any IAU General Assemblies or its professional committees' meetings. These included the 1964 Hamburg General Assembly in West Germany, the 1967 Prague General Assembly in Czechoslovakia, the 1970 Brighton General Assembly in the UK, the 1973 Sydney General Assembly in Australia, and the 1976 Grenoble General Assembly in France. This was as if a clear blue sky was spotted by dark clouds, a sky that lacked its blue color was no longer perfect. This exceedingly twisted reality of the international astronomical relations did stir up some discontent among the righteous astronomers, but it still seemed to lack the wind to blow the dark clouds away.

During the 16th IAU General Assembly in Grenoble, the IAU President Goldberg mentioned the China membership question in his opening speech. He also pointed out the talks with the PRC had reached a point of deadlock.

Goldberg mentioned "Why did China leave the IAU? Because, if I may quote from the joint statement issued at the conclusion of President Nixon's visit to Peking in 1972, the Chinese government 'firmly opposes any activities which aim at the creation of 'one China, one Taiwan', 'one China, two Governments', 'two Chinas', and 'independent Taiwan' or advocate that 'the status of Taiwan remains to be determined'. They insist that Taiwan is a province of China and therefore that the Academia Sinica of Peking must be recognized as the sole adhering organization representing all of China. Furthermore, they would require the IAU to deny admission to representatives from Taiwan to all IAU sponsored conferences. For the IAU to accept these conditions would not only violate both its principles and statutes, but would establish a precedent that could lead in the future to the politically motivated expulsion of other members of the Union."

This standstill did not breach for the many years to come. This may also have to do with what happened inside of China. Starting in 1966, the Chinese Mainland fell into a never-before-seen Great Proletarian Cultural Revolution. The values of culture were stomped upon merciless. Cultural activities such as education, sciences, and others were all put on hold and were even degenerating. Lots of scientists and scholars stepped away from their works and social ranks. Under such condition, their contacts with the international science community were rather dreams than reality. Gradually, delegates from Taiwan replaced many of the Chinese representations in the international academic communities. The Chinese Mainland's relation with the IAU also entered into the "Ice Age". When 1979 came around, within the 18 international sciences organizations, besides from the International Union of Food Science and Technology (IUFOST) and the International Union of

Immunological Societies (IUIS) for which neither Mainland China nor Taiwan participated, the Chinese Mainland had left the other 12 of the 18 organizations. They included the IAU, International Union of Pure and Applied Chemistry (IUPAC), International Union of Radio Science (URSI), International Union of Pure and Applied Physics (IUPAP), International Union of Biological Sciences (IUBS), and others. Fundamentally, the U.S. government's long-term hostile policy toward China resulted this situation.

In the end, it was better for the doer to undo what he has done. The change of global political divide in the 1970s also pushed for a de-escalation and warming up of relations between the USA and China. This began in 1971 when a warm wind blew from the United Nations. During the 26th UN General Assembly with the majority support, the UN General Assembly passed the famous Resolution 2758. This resolution was passed on October 25, 1971. It "decides to restore all its rights to the People's Republic of China and to recognize the representatives of its Government as the only legitimate representatives of China to the United Nations, and to expel forthwith the representatives of Chiang Kai-shek from the place which they unlawfully occupy at the United Nations and in all the organizations related to it."

Soon after, President Nixon visited China for the first time in 1972. As a result of his visit, the two sides issued the Shanghai Communiqué. In which, "The United States acknowledges that all Chinese on either side of the Taiwan Strait maintain there is but one China and that Taiwan is a part of China." The two sides then set up their own Liaison offices in each other's country in 1973. Two years later, President Ford visited China; this event further opened the high-level dialog between the two countries. All of these new developments on the Sino-US relations helped to serve as political foundations, in solving the predicament in international sciences relations.

The 1971 UN Resolution had caused a huge reaction among the scientists. The urge to resume the Chinese Mainland's participation was raised up daily, the Beijing sympathizers asked the IAU Executive Committee to act quickly. They wanted Chinese astronomers to rapidly reintegrate into the IAU. There was a popular saying at the time, "if the UN could expel Taiwan, Why couldn't we do the same?". This opinion was further expressed during the Grenoble General Assembly in 1976. Some who were friendly to the PRC even asked the General Assembly directly to resume Beijing's membership. The Australian radio astronomer Wilbur Norman Christiansen (1913–2007) urged six others to resume the Chinese Mainland's membership at the IAU. They also demanded the Chinese Astronomical Society become the representative for the Chinese Mainland in the IAU. They insisted the Taiwan astronomers would stay in the IAU as "China Taiwan province Astronomical Society." The then vice-President of the IAU, the French astronomer Charles Fehrenbach (1914–2008) actively lobbied for the return of the Chinese Mainland to the IAU as well. He initiated several cooperation projects with the Chinese Mainland astronomers. These strong demands and reactions had all been reflected in Goldberg's opening speech at the Grenoble General Assembly. He used half of his speech time in describing the Chinese membership question. Although Goldberg followed the decision made by the Executive Committee:

At the end of the last General Assembly in Sydney, I once stated the Executive Committee's position that, there was no other matter as urgent as resuming the PRC membership. Unfortunately, I have to say our efforts did not achieve success. Although we have tried our best to solve the deadlock that we are in along with ICSU. Until today, the Executive Committee could not find a solution where in resuming the PRC's IAU membership that will not exclude Taiwan. The removal of a member is something that is against the Statutes of the IAU.

Goldberg stated the IAU did not have interests "in the special political relationship between Taiwan and China." However, nor it "demand Taiwanese scientists to become part of China's group as a prior condition, in order for them to become members of the international sciences community." Goldberg emphasized the idea that all astronomers have the right to be represented at the IAU. This has nothing to do with any political consideration. He stated the UN chose it members based on political positions, not at the IAU. Goldberg especially underlined that when the PRC was welcomed into the IAU, it did not yet to become the representative of China at the UN. It was clear Goldberg firmly insisted on the idea that the IAU was a non-governmental academic organization. It does not take into any political elements when it accepts new members. He mentioned various proposals in solving this issue, including repeating some people suggesting to remove Taiwan's membership directly, some urged the IAU should add a clause where all member countries had to reach a certain level at their astronomical activities. Some thought they should only allow members of the UN or UNESCO to participate in the IAU's activities. Based on Goldberg's own opinion, these suggestions all violated the IAU's principles and therefore they should not be accepted. Goldberg even worried if these suggestions were implemented, some countries would leave the Union as a sign of protest, but also due to them not meeting the academic requirement. Still, he expressed that he would continue to push forward in solving the Chinese membership question.

Although there are difficulties, I do believe the current deadlock is not unsolvable. We have not only to ask China to change its political attitude towards Taiwan but also to ensure the continuing participation of Taiwanese astronomers in the IAU activities.

Goldberg encouraged astronomers from all countries to reinforce their ties and collaborations with Chinese astronomers. He declared that:

...we, astronomers should accelerate our rate of interaction with the Chinese colleagues. This can take place at the individual and the national level to take charge in development and cooperation. There are many signs that are showing China has returned to conduct and develop their astronomical activities. These signs also include their willingness in welcoming other countries' astronomers to cooperate and exchange with them. For example... upon the Chinese request, the Kitt Peak National Observatory recently sent a copy of the design paper of their two biggest telescopes to the Beijing Observatory. They asked the Chinese astronomers to participate in the joint talks of designing telescopes. I will urge astronomers... to use all opportunities to invite Chinese astronomers to attend the meetings of their home countries, to let the Chinese know how deep we are longing for their exchange and works.

The earliest meeting between Chinese and foreign astronomers happened approximately in 1974. In July that year, the International Radio Consultative Committee

(CCIR) held an international radio frequency meeting in Geneva. The Assistant General Secretary of the IAU Edith Alice Müller (1918–1995) participated in this meeting with the Chinese radio astronomy specialist Wang Shouguan. Wang worked at the Beijing Observatory at the time. The two of them politely greeted each other, although they did not get to talk about the resumption of the Chinese Mainland membership in the IAU, but this was the first occasion where a Chinese astronomer came to face to face with a IAU official since more than a decade.

Continuous developments were also taking place within the Executive Committee of the IAU. The Dutch astronomer Adriaan Blaauw became its President in 1976. This former Director of the European Southern Observatory was not only good at studying star formation, the motions of star clusters, and stellar associations but also specialized in managing all kinds of complex relations. He was the person behind the integration of various European astronomical publications and the eventual establishment of the "Astronomy and Astrophysics" journal in 1968. Blaauw also served as the first Editor-in-Chief of this journal. Since this Dutch became the President of the IAU, he clearly had faced the pressure to resume the Chinese Mainland's membership from the start. For instance, in September 1976, a document coming from Leiden titled "Legal issues regarding the ROC (Taiwan)'s membership" had reached at Blaauw's desk. It openly pointed out "the IAU's decision in accepting the ROC's application in 1959 violated its Statues, this decision was invalidate." This document further emphasized that "even if the original decision in accepting ROC's membership was legally acceptable, but to continue to allow the ROC to represent the Taiwanese people is a violation to today's reality." The legal issue to which this document refers is the amendment to the Statutes made by the IAU in 1970. It deleted the Statutes' previous description on the notion of "countries" which included "any autonomous region, protectorates, and territories that have independent astronomical activities." As a non-specialist, Blaauw had to seek the opinion of a lawyer to clarify, if Taiwan's membership was according to the current IAU Statutes. These mounting calls and pressures all made the China question become a regular item on the agenda of the Executive Committee. However, beyond this, Blaauw was excellent in dealing with international affairs. He was also fortunate enough to meet a newly matured good opportunity.

The Cultural Revolution finally ended in 1979. Chinese politics, economics, education, and sciences once again returned to their normal orbits. The Open Door policy further welcomed a robust era of development and growth. It also literally opened the door for China to have a dialog with the world on sciences. There was a mounting trend of international cultural and scientific exchanges. From 1977 to 1981, the Chinese Academy of Sciences had reached talent exchange agreements and other international cooperation agreements with scientific organizations from West Germany, France, the UK, Italy, Sweden, the USA, Japan, the Netherlands, and other countries. In November 1976, Wang Shouguan of the Beijing Observatory led a Chinese Academy of Sciences team that toured the USA. In doing so, they became the first international delegation from China to the world after the Cultural Revolution. Probably, this first Chinese Mainland astronomy delegation visited the USA for a quarter of a century. This visit again unveiled the international exchange of the astronomy field in China. Especially when Mr. Zhou Peiguang (周培源) replaced Mr. Li Siguang (李四光) to become the acting President of the China Association for Science and Technology (CAST) in 1977, the Chinese astronomy field became the "leading soldier" for China to return to the various international sciences organizations. Zhou Peiguang later became the President of the CAST in 1980. As one of the earliest members of the Chinese Astronomical Society, as well as a former part-time researcher at the Purple Mountains Observatory, thanks to Zhou's efforts and approval, from the late 1970s until the early 1980s, many countries astronomy delegations had all come to visit China. Many young Chinese astronomers were also be able to visit and study abroad. This sudden reemergence of interaction further amplified the knowledge and understanding between the Chinese and international astronomical communities. The ice was finally starting to melt. Of course, a giant iceberg cannot melt all in one day. It always takes time to solve this kind of historical dispute. For instance, the IAU was about to hold an Asia-Pacific astronomy meeting in New Zealand in 1977, the organizers of the meeting had sent out invitations to the PRC astronomers. Due to the lack of maturity of the timing, Zhang Yuzhe refused this invitation. He also once again stressed to the IAU that it should not let such contraflow of history continue over the Taiwan question. He proclaimed, "under such circumstances, we will not participate in this kind of meeting."

In both 1976 and 1977, the International Union of Geodesy and Geophysics (IUGG) and the International Union of Geological Sciences (IUGS) all passed the motion to remove Taiwan from lists of member countries. They publicly recognized "the PRC is the legitimate representative of China and Taiwan is a province of the PRC." They further claimed "only the PRC has the right to represent China, Taiwan's right of representation till this day should be canceled." These two international sciences organizations' way of resolving the Taiwan deadlock has set a precedent for others in dealing with this delicate issue. In this regard, the Chinese Academy of Sciences, the National Surveying & Mapping Bureau, and the PRC Foreign Ministry issued a joint statement in February 1977. Together, they emphasized "the eviction of Taiwan's ruling Chiang and under no circumstances, there should appear 'two Chinas', 'one China, one Taiwan' as membership requirements." In regards to Taiwanese scientists' individual participation in in international meetings, this statement emphasized so long they "would not be branded as from the ROC, or they did not enter the meetings in the name of any of the Chiang Kaï-Shek's nationalist official organs," the PRC side would not interfere in Taiwanese scientists' individual attendance.

In 1978, the International Radio Consultative Committee held another meeting in Geneva. The then General Secretary of the IAU Edith Alice Müller received an invitation by the PRC delegation. Both sides then discussed the possibility of resuming the PRC's membership at the IAU during that meeting. Although their discussion was not much different from previous interactions, this event helped them further to understand the opposite side's intentions and positions. After the meeting with the PRC delegation, Müller wrote a letter to Blaauw, in which she pointed out the only difference from the Chinese position, was "they adamantly required the IAU to follow the footsteps of the IUGG and IUGS." Sure enough, the IAU Executive Committee once again rejected this proposition. Several months later, with the invitation from the Dutch Academy of Sciences, the Chinese Academy of Sciences delegation visited the Netherlands. During this visit, the Chinese officials had an extensive opportunity to further discuss this matter with the IAU officials. That was because the President of the IAU at the time Adriaan Blaauw was from the Netherlands. He was also a professor at Leiden University. While the Chinese were visiting the Leiden Observatory on July 13, Blaauw gave a speech that included contents about the European Southern Observatory, astronomy, astrophysics, and the IAU's international relations. His message behind this speech was obvious in the Chinese delegation's eyes. During the various occasions following suit, including a banquet held at the Chinese Embassy in the Netherlands, Blaauw exchanged his point of view with Chinese officials, including Zhu Yongxing (朱永行, 1930-), who was the Deputy Bureau Chief of the Chinese Academy of Sciences Foreign Affairs Bureau. Blaauw was left with the impression that the Chinese officials asked the IAU to wait for the right opportunity and the de-escalation of both sides' opposition of views. In early 1979, the IAU Executive Committee reached a consensus. They wanted both the Chinese Mainland and Taiwan's astronomers could be able to participate in its activities. They also expected to make certain concessions with the PRC's side.

The situation was quickly evolving. The PRC's policy in regards to its participation in international organizations changed ingeniously as well. Some policies were also adjusted in Beijing. Under the new policy guidelines, the PRC's participation in international organizations no longer firmly demands the eviction of the Taiwan delegation. They only stressed the idea that Beijing would not accept "two Chinas" or "one China, one Taiwan." This series of signals left fresh impressions on Blaauw. It also regenerated the IAU Executive Committee's hope and confidence in finally solving the China and Taiwan dilemma. It was time for them to take action.

To let the Chinese Mainland, Taiwan and the IAU's representatives meet directly was the best and the most direct solution in solving this obstacle. The Montreal General Assembly would soon arrive on August 14, 1979. It was also the 17th IAU General Assembly. This was a God-given chance in solving this issue. However, was there any willingness from the Chinese Mainland or Taiwan? The Executive Committee then proceed to test the waters at both sides of the Taiwan Strait.

On April 9, 1979, Blaauw wrote a letter to Academia Sinica in Taiwan. This letter not only invited Taiwan's delegation to attend the Montreal Assembly but not also proposed three formulas in ironing out the Taiwan question. There were the following proposed by Blaauw: (1) the Chinese Academy of Sciences in Beijing and Academia Sinica in Taipei could form a joint delegation to represent all Chinese Scientists. (2) Both sides could participate in the IAU on a voluntary basis, Taiwan giving up its voting rights in General Assemblies and in the Finance Committee. (3) The Chinese Mainland could reach an agreement with Taiwan, so that two sides would only send out delegates as individual members to participate in the IAU activities.

The one who sent out the invitation to the Chinese Mainland was the then Assistant General Secretary of the IAU, the Irish astronomer Patrick Arthur Wayman (1927–1998). He served as the Director of the Dunsink Observatory near Dublin at the time. He went to China partly because he had already been scheduled to give lectures in April that year in the Chinese Mainland. Wayman thus had the chance to have unofficial discussions with Chinese officials and astronomers during this trip. On April 23, Wayman had a talk with Wang Shouguan, Hong Siyi (1926-), and the Foreign Affairs Bureau Chief of the China Association for Science and Technology Fang Jun (方均) among others. Wayman forwarded to them the three solutions mentioned above. After the discussion ended, both sides signed a memorandum. It stated that both the IAU and the PRC would consider the first solution; the second and third solutions were not into consideration. At the same time, Blaauw and others had received the answer from Taipei: Taiwan did not specifically indicate its preference. They only agreed to continue the dialog and they invited the IAU to send officials to visit Taiwan. They wanted them to learn more about the astronomical activities there. The Taipei letter stressed that the Taiwan Astronomical Society would not represent Mainland astronomers, and they were not against the resumption of the PRC delegation on the condition that the PRC would not represent Taiwanese scientists. When they received the answer from Taiwan, the Montreal General Assembly was getting close, therefore the IAU officials could not organize this visit.

Blaauw had grasped the accurate picture. The Chinese Mainland's policy toward its integration into non-governmental international sciences organizations had indeed loosened and relaxed. On June 11, 1979, the Chinese Academy of Sciences, the China Association for Science and Technology, the State Scientific and Technological Commission, the Chinese Foreign Ministry made a joint plea to the State Council of the PRC regarding the membership of some of the ICSU member unions. Together, they suggested to the PRC central government that "if the International Union of Biochemistry and Molecular Biology (IUBMB) and the IAU officially demand us to form a joint organization with Taiwan to participate in their activities, we will concur to this suggestion." Evidently, Wayman's first solution was reported to the Ministry of Science and Technology and Foreign Affairs of the PRC. The high-level officials agreed to this proposal. This joint report also asked permission to send five to six delegates to attend the 11th IUBMB General Assembly and the 17th IAU General Assembly in Montreal. It also asked permission from the Central Government to send delegates to attend the IUPAC General Assembly in Helsinki and the 24th International Geographical Congress in Tokyo in 1980. This document opened the door for the participation of the 17th IAU General Assembly in Montreal for both the Chinese Mainland and Taiwan. It paved the foundation for the Mainland's return to the IAU.

The opportunity had finally matured; it was the time to make major breakthroughs. Since the scientists from the Chinese Mainland and Taiwan had a clear willingness in participating to the Montreal Assembly, Blaauw sent out two invitations in the name of the President of the IAU. He asked both sides to send astronomers to this Assembly as "invited delegates." On July 29, 1979, Beijing's and Taipei's answers had returned to the IAU, they both accepted the invitation. They also expressed their will to have talks at the Montreal Assembly. For the Chinese Mainland's side, the head of the delegation was Zhang Yuzhe, who was the long-time President of the Chinese Astronomical Society. His deputy was Zhao Xianzi (1926–1996), who was from the Purple Mountain Observatory. Four other members went to Montreal in August with them. They included Ye Shuhua (1927-), who worked at the Shanghai Observatory, Hong Siyi of the Beijing Observatory, Yi Zhaohua of the Nanjing University (1932–2017), and Zhu Jinning who was an official from the China Association for Science and Technology. On the Taiwan side, Shen Junshan (沈君山, 1932–2018) was the only delegate. He was the then President of the Taiwan Astronomical Society.

The two sides started their talks from August 13 until 24, which was longer than the Assembly agenda itself. It was a friendly but laborious discussion. The Mainland side wore a name tag on which was written "China: Nanjing," and Shen Junshan wore his which stated "China: Taipei" on his chest. Since the Chinese Mainland was yet to return to the IAU, so its delegation did not participate in the General Assembly, but took part in the sideline academic discussions. On one side, they did so to display the strength of the Chinese Mainland's astronomy field, on the other side it was also to prevent possible misunderstandings of certain results having been reached at the General Assembly.

As the sole Taiwan representative to the Montreal Assembly, Shen Junshan was born in Nanjing; his paternal family roots are in Yuyao, Zhejiang province. His father Shen Tsung-han (沈宗瀚) was an agricultural expert. Shen's parents were both highly educated and had studied in the USA; his maternal grandfather also went to France as an exchange student. Shen followed his father to Taiwan a few years later 1949. Shen's father rose to further political prominence in Taiwan, eventually becoming the Director of Sino-American Joint Commission on Rural Reconstruction. Shen Junshan himself graduated from National Taiwan University's physics department in 1955. In 1957, he left Taiwan for the USA, to enroll in a doctoral program in physics at the University of Maryland. He graduated from there in 1961. Shen's dissertation was entitled "Application of a Dispersion Relation to the Electron Impact Widths and Shifts of Isolated Spectral Lines from Neutral Atoms." Shen then worked at Princeton University and Purdue University as well as took up a position at NASA before returning to Taiwan. He returned to Taiwan in 1973 to take up a post as the head of National Tsing Hua University's sciences faculty, at a salary only one-eighth that which he received in the United States, earning him praise as a "model of patriotism" for his actions. Since his father occupied prominent role in the ROC government in Taiwan, Shen and three other youngsters at the time were given the nickname of the "Four Princelings" between 1970s–1990s. Shen Junshan was not only a reputed scholar in astronomy but also a cultured fellow in Classical Chinese arts. He was a renaissance man. His special status made him a perfect candidate to manage the political debate in Taiwan, as well as handle the Cross-Strait relations. In October 1979, the agreement between both sides in regards to their adhesions to the International Olympics Committee would also be a result of Shen Junshan's doing. He represented Taiwan as a member of the ROC Olympics Committee and discussed this matter with the PRC Olympics Committee officials in Nagoya, Japan. From this meeting, the Mainland and Taiwan delegation agreed Beijing's participation in the International Olympic Committee. Taiwan would use the title "Chinese Taipei" to continue to participate in Olympic events. This formula has been kept as a model in solving the Taiwan question for many international organizations to come. After the Montreal Assembly, Shen even visited the Main-land on multiple occasions; he contributed greatly to the promotion of Cross-Strait academic cooperation and developments. That is a later story.

The most challenging obstacle was both sides' political positions. However, after a series of sincere exchanges, they all welcomed each other in participating in the IAU activities. This remained as the basis of their talks. During this time, Blaauw surprisingly (?) and joyfully found out that Zhang Yuzhe was a former colleague of his some 30 years ago. Zhang and Blaauw both worked at the Yerkes Observatory under its then Director Otto Struve. These shared memories between the two of them also facilitated their demanding talks.

Shen Junshan treated the Mainland delegation with politeness and respect; he called Zhang Yuzhe his "senpai"—his "elder." The talks included many aspects and both sides were clear: No matter what the topic was, they were all under the principle that China was indivisible. Any "two Chinas," "one China, one Taiwan" suggestions were not permitted. Near the end of their talks, the two sides focused on the conflict of finding a suitable name for Taiwan's delegation that would be not only in line with the above-mentioned principles but also a name which could be agreed upon by both sides. The Mainland suggested names such as "China Taiwan Astronomical Soci-ety," or "China Taiwan Astronomy," but Shen Junshan rejected them all. This deadlock frustrated Blaauw and he kept on walking back and forth in his office while trying to find a better name for the Taiwan delegation. He even sighed that "this was totally incomprehensible; these ideas which seemed to be good yet were still rejected!" Blaauw's frustration might be a result of Western-Eastern cultural differences. Shen Junshan was really preoccupied, for lack of help. He called Taipei nearly every day to forward the talk contents and ask for permission. Even so, the three sides at the talk table could not find a suitable name for Taiwan to remain at the IAU that was acceptable by everyone.

In the end, the Chinese Mainland delegation suggested to suspend the Taiwan delegation's name debate. This was to be treated in later times. In the beginning, the IAU Executive Committee rejected this proposal. Nonetheless, as the Montreal Gen-eral Assembly was reaching its end, the three sides quickly realized that if they could not reach a consensus through these talks, the Taiwan question would remain unsettled afterward. Things could be in trouble again. Based on this understanding, as well as the PRC delegation's relentless efforts, the Executive Committee finally adopted the temporary suspension of the question of the Taiwan delegation's name. Since the Taiwanese side also did not yet approve this proposal, the Executive Committee further adopted a compromising strategy, where the agreement was not to be submitted to the General Assembly but to proceed by a letter exchange. In doing so, it avoided the possibility of lack of support in the General Assembly's voting process. This truly particular agreement was called "the Montreal Agreement."

The Montreal Agreement was then put forward to the Assembly as a letter exchange between Zhang Yuzhe and Adriaan Blaauw. Its signing date was on August 22, 1979. The entirety of these letters was published at the IAU's General

Assembly's 17th publication Appendix IV., on Volume B's page 526. *These two English letters' are the following:*

526

Montreal, August 22, 1979

To Professor A. BLAAUW, President
of the International Astronomical Union
c/o the General Assembly of the IAU 1979
Montreal, Quebec, Canada

Dear Professor Blaauw,

On behalf of the Chinese Astronomical Society, whose delegates have convened with IAU Officers during the 17th General Assembly of IAU at Montreal, I herewith confirm that,

considering that both the Chinese Astronomical Society and the IAU believe that the question of China's representation in the IAU needs urgent attention;

noting with satisfaction that the IAU wishes to find a solution to the question on the principle of one China, embracing the Mainland and Taiwan;

understanding the nature of complexity of such an endeavor in the present circumstances;

considering also the desire expressed by the IAU that the Chinese Astronomical Society is welcome to return without preventing the astronomical organization as well as astronomers in Taiwan from participating in the IAU;

reiterating that the Chinese Astronomical Society opposes any solutions which can suggest the existence of "two Chinas" or "one China, one Taiwan";

the Chinese Astronomical Society proposes to the IAU Executive Committee

   a) that for the time being dual adherence should be introduced, the adherence and the adhering bodies to be specified as:

China ⟨
   — Chinese Astronomical Society (Purple Mountain Observatory, Nanking)
   — (a name for the adhering organization representing the astronomical community of Taiwan; for this name we propose "Chinese Taiwan Astronomical Society" or, provisionally, as reference "Astronomy Taiwan, China"; the name to be established by the IAU Executive Committee in consultation with the two adhering organizations);

   b) that for the time being the financial contributions of China be composed of one component due from the Chinese Astronomical Society (Nanking) and one component due from (the name to be defined as under (a));

   c) that for the time being each of these adhering bodies nominates a representative authorized to vote in the General Assembly. In accordance with article VII, 11 of the IAU Statutes, on questions

527

involving the budget of the Union, the number of votes of these
representatives will be one greater than the number of its category
as defined in article 8, and on questions of administration, not
involving the budget, each representative will have one vote.

Yours sincerely,

Y.C. Chang, President
Chinese Astronomical Society

August 22, 1979

To Prof. Y.C. CHANG, President
of the Chinese Astronomical Society
c/o IAU General Assembly 1979
Montreal, Canada

Dear Professor Chang,

In reply to your letter of August 22, 1979 I have the pleasure to inform
you as follows:

The Executive Committee has taken note with great satisfaction of the
proposal contained in your letter, recognizing the Chinese Astronomical
Society's view leading to this proposal. It wishes to express its high
appreciation for the constructive spirit in which the delegation of your
Society has contributed towards restoring full participation of China in the
IAU.

The Executive Committee therefore will be glad to take up the problem of
deciding on the name to be used for the adhering body on behalf of the
astronomical community of Taiwan. It will do so in the near future in
consultation with the astronomical community of Taiwan and with your Society,
and possibly benefiting also from the experience gained meanwhile by other
scientific unions working similarly towards the full participation of China.

The Executive Committee, in view of the discussions held during this
Assembly, hopes and expects that this course will soon lead to the formal
restoration of full Chinese participation and would be extremely pleased to
see the valuable participation of members of your Society in the scientific
discussions at this Assembly, continued at the forthcoming scientific meetings
of the IAU. To this purpose, the General Secretariat will supply your Society
with all relevant documentation.

Yours respectfully,

A. Blaauw, President
International Astronomical Union

In this letter, Zhang Yuzhe used the word "temporary." This symbolized the fact
that the Chinese Astronomical Society had the wishes and the position for the
unification of China in the end.

On one side, it was to achieve some kind of balance; on the other, it was to give
time to the Taiwanese side to state their position. On August 23, during the second
plenary meeting of the Montreal General Assembly, after Adriaan Blaauw had

spoken, he asked if Shen Junshan also wanted to speak a word or two during this meeting. Shen took over and stated that "the Taiwanese astronomers and the government of Taiwan do not oppose the return of the Chinese Mainland to the IAU." However, he argued as the Chinese Mainland insisted that Taiwan was a province of the PRC, thus "the Nanjing (astronomy) delegation was a national representation, and the Taipei (astronomy) delegation was, therefore, a regional entity." Shen stated that they "would not accept to enter the IAU as a regional representation," because they represented "an independent academic realm." Therefore, Shen Junshan claimed that, "before the next General Assembly, our fellow academic realm from Nanjing could join us on the basis of respect for the principle of reciprocity of membership."

Division of opinions still reigned the day, the ice still needed time to melt. Nevertheless, during Shen's speech he clearly mentioned the "The Island of Taiwan," "two parts of China." This expressed Taiwan's recognition of the "one China" principle, and opened the way to solve the remaining problems.

The Montreal General Assembly not only reached the above-mentioned Agreement, but the Executive Committee also put forward a draft resolution to the General Assembly. During its second plenary meeting, President Blaauw reported the developments of the negotiation to the General Assembly. He also read out the following resolution:

"The General Assembly,

in view of the reported progress in restoring full adherence of China;

noting the letters on this subject exchanged between the President of the IAU and the President of the Chinese Astronomical Society, Prof. Y.C. Chang;

authorizes the Executive Committee to implement the anticipated conclusion of the current negotiations subject to ratification by the General Assembly at its next meeting."

This resolution then was submitted to the General Assembly and carried unanimously.

The Montreal General Assembly passed this resolution. It was then fully published on page 49, Volume 17 of the IAU's General Assembly publication that year. Still, the full reinstatement of China's membership at the IAU was not fully resolved at the Montreal General Assembly. Taiwan's membership name became the only remaining question from that General Assembly. Yet, this was not a strange outcome, because it involved fundamental political and principle questions. Time was necessary for the two sides to consider and to accept it. Blaauw believed the IAU's way of solving the Taiwan question could serve as a precedent for other international scientific unions. At the Montreal General Assembly, he confidently predicted that the rest of the problem would eventually find its solution by the end of that year.

The Chinese Mainland finally entirely resolved its membership issue at the IAU in March 1980. In 1982, the relevant resolution was passed during the 18th General Assembly in Patras, Greece. The following were the designated information for both Mainland China and Taiwan on page 4214, Volume 18 of the IAU General Assembly publication that year.

```
    It was recommended that adherence of China be now as agreed with autho-
rities in both Nanjing (formerly Nanking) and Taipei as :

                              Chinese Astronomical Society
                              Purple Mountain Observatory
                              Nanjing, China

                  and :
                              Chinese Astronomy Union
                              Taipei Observatory
                              Taipei, Taiwan
                              China
```

The first plenary meeting of the Patras General Assembly was held on August 17 at the Roman Odeon there. It is in the upper town. The Odeon was built around 160 AD, during the reign of either Antoninus Pius or Marcus Aurelius. It had been restored and partially reconstructed and is used as an open-air theater for performances and concerts during the summer months. From there, one can have a bird's-eye view of this abandoned old city. This was the result of the wounds from the Venetian invasion some 300 years ago. Today, this ancient theater had welcomed guests from another ancient civilization, the Chinese Mainland astronomy delegation. The fourth item of the agenda from the first plenary meeting was for the General Secretary to give the Executive Committee's first half-year report in 1982. The full reinstatement of China's membership at the IAU was among its main contents.

General Secretary Patrick A. Wayman gave a detailed speech on the Executive Committee's advice on permitting the reinstatement of China's membership in the IAU. He mentioned that the 1979 General Assembly had already granted authorization to the Executive Committee to carry out the Montreal Talks. The problem was basically solved then and there. Since May 1980, the Chinese Astronomical Society in Nanjing had thoroughly reached its membership requirements. These included affirming individual IAU members, nominating the professional commission members, and providing financial assistance to the Mainland astronomers. These also included the Chinese participation at the International Time Bureau among other tasks. In regards to Taiwan's name question at the IAU membership list, it was also solved on May 1, 1980. The new name for the Taiwanese delegation not only satisfied the Chinese Astronomical Society but also received approval from Taipei. Therefore, Wayman asked all delegations of the General Assembly to pass the Executive Committee's action in fully confirming China's membership at the IAU, following the Executive Committee's previous decision at the 17th General Assembly. The action that Wayman had referred to was the A1 resolution adopted by the IAU General Assembly in 1982. Its contents are the following:

Resolution No A 1
Ratification of IAU Membership of China
Ratification de l'adhésion de la Chine à l'UAI

Ratification of IAU Membership of China

The General Assembly

noting

that progress has been achieved in restoring full adherence to the IAU of China, as was anticipated at the XVIIth General Assembly of 1979 at Montreal, and that agreement was obtained by 1 May 1980 on the mode of listing, in the official list of member countries, two adhering bodies for China as a temporary measure,

ratifies

the arrangements made by the Executive Committee for the adherence of China to the Union during the period elapsed since the XVIIth General Assembly.

The General Assembly passed this resolution unanimously under applause. Soon after, the executive President of the General Assembly invited the head of the Mainland China delegation Wang Shouguan to make a few remarks on the behalf of the Chinese Astronomical Society. He made the following speech:

Dear friends & colleagues,

The Chinese Astronomical Society celebrates its 60th anniversary this year. Its reunion with this international community today is an event that is highly appreciated by all its 900 members. I and my colleagues here are very glad to have this opportunity of speaking on behalf of our Society and its members to express our most cordial greetings and most sincere thanks to you all. Thank you!

Wang's speech was welcomed by the delegates' a warm round of applause.

Although his concise speech only included close to one hundred Chinese characters, it represented members of the Chinese Astronomical Society's aspiration to return to the IAU, as well as both the process and the happy ending that the Society was able to do so.

Even though the Chinese Mainland delegation was able to achieve their primary objective, but they ran into a brief interlude soon after their arrival in Patras. What happened was, when the Chinese Mainland delegation got their hands on the agenda of the General Assembly, they found out some of the wordings were clearly different from what was concurred in the agreement. There was a problem of principle. The head of the delegation Wang Shouguan then reached out to Wayman for an explanation but did not receive a satisfactory answer. Therefore, he went to seek help from the retired President Blaauw and explained the magnitude of the problem to him.

The next day, Blaauw went over to the Chinese Mainland delegation, he agreed that there were inconveniences included in the agenda. He further suggested changing the agenda based on the previous agreement and invited Wang Shouguan to make a speech during the General Assembly.

China's legitimate seat at the IAU was finally entirely restored. 93 astronomers from the Mainland and 16 Chinese astronomers from Taiwan became the first batch of its members. From that day on, Chinese astronomers have gradually become a consequential team in the international astronomical arena. We have shown off our talents and capacities to the world.

References

IAU: General Web Resources

Directory: Individual members https://www.iau.org/administration/membership/individual/
Executive Committees: Past https://www.iau.org/administration/executive_bodies/past_commit tees/executive_committee/
Executive Committees: Minutes https://www.iau.org/administration/executive_bodies/executive_ committee/ecminutes/
General Assemblies. https://www.iau.org/science/meetings/past/General_assemblies/
General Assemblies: Proceedings ("Transactions B") https://www.iau.org/publications/iau/trans actions_b/
General Assemblies: Daily Newspapers https://www.iau.org/publications/iau/ga_newspapers/
Symposia: Proceedings https://www.iau.org/publications/iau/symposia/

"IAU Transactions"[37]

Beijing (GA XXVIII, 2012): Transactions Vol.XXVIII B, Ed. T. Montmerle, Cambridge University Press (2015).
Berkeley (GA XI, 1961): Transactions Vol.XI B, Ed. D.H. Sadler, Blackwell Scientific Publications (1962).[38]

---

[37] *In full:* "IAU Transactions [Vol. x], Proceedings of the International Astronomical Union".
    Until and including the Moscow General Assembly in 1958, only one volume of the Transactions was published, edited by the General Secretary elected at that Assembly. As of the Berkeley General Assembly in 1961, two volumes were published, separating the Scientific part (vol. "A") from the Administrative part (vol. "B"). Vol. A is edited by the incoming General Secretary, vol. B by the outgoing General Secretary. In the present text, only the Transactions "B" are relevant after 1961 (see the Internet link above). The volumes are referred to by "GA" (= General Assembly).
[38] Due to outstanding copyright issues, this volume is unfortunately not yet available online.

T. Montmerle, Y. Zhou, *China and the International Astronomical Union*, Historical & Cultural Astronomy, https://doi.org/10.1007/978-3-031-01787-2

Brighton (GA XIV, 1970): Transactions Vol.XIV B, Ed. C. de Jager & A. Jappel, Association of Universities for Research in Astronomy (AURA) (1971).

Dublin (GA IX, 1955): Transactions Vol.IX, Ed. P. Th. Oosterhoff, Cambridge University Press (1957).

Grenoble (GA XVI, 1976): Transactions Vol.XVI B, Ed. E.A. Müller & A. Jappel, Association of Universities for Research in Astronomy (AURA) (1977).

Montreal (GA XVII, 1979): Transactions Vol.XVII B, Ed. P. Wayman, Association of Universities for Research in Astronomy (AURA) (1980).

Moscow (GA X, 1958): Transactions Vol.X, Ed. D.H. Sadler, Cambridge University Press (1960).

Paris (GA V, 1935): Transactions Vol.V, Ed. F.J.M. Stratton, Cambridge University Press (1936).

Patras (GA XVIII, 1982): Transactions Vol.XVIII B, Ed. R. West, Association of Universities for Research in Astronomy (AURA) (1983).

Stockholm (GA VI, 1938): Transactions Vol.VI, Ed. J.H. Oort, Cambridge University Press (1939).

Sydney (GA XV, 1973): Transactions Vol.XV B, Ed. G. Contopoulos & A. Jappel, Association of Universities for Research in Astronomy (AURA) (1974).

Books & Reports

Andersen, J., Baneke, D., & Madsen, C. (2019). *The International Astronomical Union, uniting the community for 100 years* (376 pp.). Springer Nature Switzerland.

Appenzeller, I., Chmielewsky, Y., Pecker, J.-C., de la Reza, R., Tammann, G., & Wayman, P. (Eds.). (1998). *Remembering Edith Alice Müller* (156 pp.). Astrophysics and Space Science Library, Kluwer Academic Publishers.

Blaauw, A. (1994). *History of the IAU, the birth and first half-century of the International Astronomical Union* (296 pp.). Kluwer Academic Publishers.

Cuming, B. (2010). *The Korean war: A history* (290 pp.). The Modern Library.

Espy, R. (1979). *The politics of the Olympic games* (p. 212). The University of California Press.

Fehrenbach, Ch. (1990). *Des hommes, des télescopes, des étoiles* (257 pp.). Editions du CNRS). [in French].

Fu, C., & Ye, S. (2009). *Under One Starry Sky* 《同一个星空:国际天文学联合会史》 – *History of the International Astronomical Union* (pp. 128–167). [In Chinese: Shanghai Jiao Tong University Press, Shanghai].

Goldberg, L., & Edwards, L. (Eds.) (1979). *Astronomy in China,* CSCPRC report no. 7 (110 pp.). National Academy of Sciences.

Greenaway, F. (1996). *A history of the International Council of Scientific Unions* (280 pp.). Cambridge University Press.

Montmerle, T., & Fauque, D. (Eds.). (2022). *Astronomers as diplomats: When the IAU builds bridges between nations* (Historical and cultural astronomy series). Springer Nature Switzerland. in press.

Montmerle, T., Zhou, Y., & Gomas, Y. (2022). [MZG22], *The two lives of Cheng Maolan: From the French "silk road towards astronomy" to the meanders of Mao's China.* Springer Nature Switzerland, in press.

Olson, L. (2014). *Those angry days: Roosevelt, Lingbergh, and America's fight over World War II (1939–1941)* (548 pp.). Random House.

Plating, J. D. (2011). *The Hump: American strategy for keeping China in World War II.* (331 pp.). Texas A&M University Press.

Roux, A. (2016). *Chiang Kaï-shek, Le grand rival de Mao* (654 pp.). Editions Payot & Rivages. [in French].

Shen, C.-S. (2004). *The postscript of the floating life: One, but not by unification (*浮生後記: 一而不統). Global Views – Commonwealth Publishing.

Short, P. (1999). *Mao: A life* (842 pp.). Hodder & Stoughton.

Sterken, C., Hearnshaw, J., & Valls-Gabaud, D. (Eds.). (2019). [IAUS 349], Under one sky: *The IAU centenary symposium, IAU symposium 349* (542 pp.). Cambridge University Press.

Articles

Aller, L. H. (1997). *Leo Goldberg (1913–1987), biographical memoirs* (Vol. 72). National Academy of Sciences.

Blaauw, A., & Schmidt, M. (1993). Jan Hendrik Oort (1900-1992). *Publications of the Astronomical Society of the Pacific, 105*, 681.

Chinnici, I. (2022). Astronomers as diplomats. In T. Montmerle & D. Fauque (Eds.), *Historical & cultural astronomy*. Springer. Chap. 1.

Fauque, D., & Fox, R. (2022). Astronomers as diplomats. In T. Montmerle & D. Fauque (Eds.), *Historical & cultural astronomy*. Springer. Chap. 3.

Hales, A. L. (1992). *Lloyd Viel Berkner (1905–1967), a biographical memoir*. National Academy of Sciences.

Lee, H. M. (2022). Astronomers as diplomats. In T. Montmerle & D. Fauque (Eds.), *Historical & cultural astronomy*. Springer. Chap. 10.

Liu, X. (2019). The China crisis. *IAUS, 349*, 222–227.

Liu, X. (2022). Astronomers as diplomats. In T. Montmerle & D. Fauque (Eds.), *Historical & cultural astronomy*. Springer. Chap. 5.

McClure, D. S. (2002). *Wallace Reed Brode (1900–1974), biographical memoirs* (Vol. 82). National Academy of Sciences.

Miley, G. K. (1974, March). Astronomy in China today. *Sky & Telescope, 47*, 148–152.

Miley, G. K. (2022). Astronomers as diplomats. In T. Montmerle & D. Fauque (Eds.), *Historical & cultural astronomy*. Springer. Chap. 12.

Montmerle, T. (2019). The IAU, from commissions to divisions... to commissions, [IAUS349], pp. 289–324.

Montmerle, T. (2022a). Astronomers as diplomats. In T. Montmerle & D. Fauque (Eds.), *Historical & cultural astronomy*. Springer. Chap. 6.

Montmerle, T. (2022b). Astronomers as diplomats. In T. Montmerle & D. Fauque (Eds.), *Historical & cultural astronomy*. Springer. Chap. 16.

Pecker, J.-C. (2019). *With the IAU and inside the IAU since 1946*, [IAUS349], pp. 112–119.

Sawyer, R. A. (1974, November). Wallace Brode. *Physics Today, 27*(11), 79.

Shouguan, W. (2009). Personal recollections of W.N. Christiansen and the early days of Chinese radioastronomy. *Journal of Astronomical History and Heritage, 12*(1), 33–38.

Shouguan Wang [see Wang Shouguan]. (2016). Reminiscence of my sixty-five year voyage in astronomy. *Research in Astronomy and Astrophysics, 16*, 6, 86. (12pp).

Shouguan, W. (2022). Astronomers as diplomats. In T. Montmerle & D. Fauque (Eds.), *Historical & cultural astronomy*. Springer. Chap. 8.

UNESCO. (2006). UNESCO and ICSU: Sixty years of cooperation. In *Sixty years of science at UNESCO 1945–2005* (pp. 496–506). On line at: https://unesdoc.unesco.org/ark:/48223/pf0000149231?29=null&queryId=3b13f6be-39ca-4fc1-ad08-fa0cf3eb8469

van de Hulst, H. C. (1994). Jan Hendrik Oort (28 April 1900–5 November 1992). *Biographical Memoirs of Fellows of the Royal Society, 40*, 320–326.

Wang, S.-Y., Wei, Y.-C., & Chou, M.-Y. (2022). Astronomers as diplomats. In T. Montmerle & D. Fauque (Eds.), *Historical & cultural astronomy*. Springer. Chap. 9.

Wielen, R. (2019). *Germany's difficulties in becoming a member of the International Astronomical Union,* IAUS349, pp. 205–213.

Wielen, R. (2022). Astronomers as diplomats. In T. Montmerle & D. Fauque (Eds.), *Historical & cultural astronomy*. Springer. Chap. 11.

Wilford, J. N. (1992, Nov 12). Jan H. Oort, Dutch astronomer in forefront of field, dies at 92. *The New York Times*. [https://nyti.ms/29bhGYj].

Wilkins, G. A. (1991). Donald Harry Sadler, O.B.E. (1908-1987). *Quarterly Journal of the Royal Astronomical Society, 32*, 59–65.

Wilkins, G. A. (Ed.) (2008). *A personal history of H.M. Nautical Almanac Office (1930–1972), by Donald H. Sadler,* HMNAO's website [http://www.hmnao.com/history].

Zhou, Y. (2022). Astronomers as diplomats. In T. Montmerle & D. Fauque (Eds.), *Historical & cultural astronomy*. Springer. Chap. 7.

Name Index

© The Author(s), under exclusive license to Springer Nature Switzerland AG 2022   209
T. Montmerle, Y. Zhou, *China and the International Astronomical Union*, Historical
& Cultural Astronomy, https://doi.org/10.1007/978-3-031-01787-2